SpringerBriefs in Applied Sciences and Technology

Multiphase Flow

Series editors

Lixin Cheng, Portsmouth, UK
Dieter Mewes, Hannover, Germany

For further volumes:
http://www.springer.com/series/11897

Shwin-Chung Wong

The Evaporation Mechanism in the Wick of Copper Heat Pipes

 Springer

Shwin-Chung Wong
Department of Power Mechanical
  Engineering
National Tsing Hua University
Hsin-Chu
Taiwan

ISSN 2191-530X   ISSN 2191-5318   (electronic)
ISBN 978-3-319-04494-1   ISBN 978-3-319-04495-8   (eBook)
DOI 10.1007/978-3-319-04495-8
Springer Cham Heidelberg New York Dordrecht London

Library of Congress Control Number: 2014930578

Springer is part of Springer Science+Business Media (www.springer.com)

Acknowledgments

This work was funded by National Science Council, R.O.C. under Contracts NSC96-2221-E-007-059-MY2 and NSC99-2221-E-007-029-MY2.

Contents

Nomenclature

A_e	Evaporator area [cm^2]
A_w	Cross-sectional area of the wick
K	Permeability of the wick
k	Thermal conductivity [W/mK]
L	Liquid flow length
N	Figure of merit [W/m^2]
Q	Net heat load [W]
Q_{cond}	Conduction through base plate [W]
Q_t	Total heat load [W]
q	Heat flux [W/cm^2]
R_e	Evaporator resistance [K/W]
r_{eff}	Effective radius of the wick pore
T	Temperature [°C]
ΔT_{sup}	Superheat at wick bottom [°C]
v	Volumetric fluid charge [ml]

Greek Symbols

δ	Wetted-wick thickness
λ	Heat of evaporation [kJ/kg]
μ	Viscosity [cP]
θ	Apparent contact angle under evaporation (°)
θ_0	Static contact angle (°)
ρ	Density [kg/m^3]
σ	Surface tension [N/m]
v_{max}	Volume flow rate under Q_{max}

Subscripts

a	Acetone
e	Evaporator
eff	Effective
l	Liquid
m	Methanol
max	Maximum
min	Minimum
w	Water, wick

Abstract

The Evaporation Mechanism in the Wick of Copper Heat Pipes are discussed based on recent visualization experiments for operating horizontal flat-plate heat pipes. First, using water as the working fluid, different wick types including sintered multilayer mesh, powder, and composite powder-mesh wick, were tested with respect to various charges, wick combinations, and wick thicknesses. Then, evaporative characteristics for water, methanol, and acetone, which have different figures of merit, were compared under various conditions of copper surface wettability using a two-layer mesh wick. Deeper knowledge was acquired by correlating the evaporator resistance data with the visualized evaporating surface and evaporation behavior. In general, when the heat load was increased, the evaporating liquid layer receded to yield decreased evaporator resistances. Beyond a maximum heat load, local dryout began at the front end of the evaporation zone accompanied by an abrupt increase of the evaporator resistance. The maximum heat loads for water were much larger than methanol and acetone, in good agreement with the sequence of their figures of merit. The evaporator resistances were lowest for water and highest for acetone, mainly related with their different average evaporating liquid layer thicknesses. Degraded surface wettability resulted in significantly lowered maximum heat loads for water, with those of methanol and acetone nearly unaffected. The evaporator resistances were hardly affected for all three fluids. Quiescent surface evaporation without nucleate boiling was observed for water under all test conditions. Very weak local nucleation was sometimes observed for methanol near the maximum heat loads. For acetone, slow, periodic up-and-down motion of the liquid layer as a result of nucleate boiling was observed near the maximum heat loads over a significant portion of the evaporator.

Keywords Heat pipe · Evaporator resistance · Evaporation heat transfer · Two-phase heat transfer · Mesh wick · Powder wick · Figure of merit · Nucleate boiling

The Evaporation Mechanism in the Wick of Copper Heat Pipes

1 Introduction

The heat pipe has been widely used for cooling high power density devices due to its high effective conductivity, good heat spreading capability and geometric flexibility. The heat from the heat source is absorbed by the working fluid at the evaporation section via evaporation. The vapor then condenses at the distant condensation section where the latent heat is released to the environment. The condensed fluid is drawn back to the evaporator through the wick by the capillary force to complete the thermal cycle. A good heat pipe is characterized by a low thermal resistance and a high dryout tolerance. The evaporator resistance is the critical part of the overall thermal resistance of a heat pipe. It is influenced by many parameters, including the heat load, the type, thickness, porosity, and permeability of the wick, the amount and properties of the working fluid, and the wick capillarity, etc. Experiments to simultaneously acquire visualization and evaporator resistances for operating heat pipes are very difficult because hermetic sealing of the heat pipe container is crucial to prevent inward leakage of non-condensable gases. Consequently, most related visualization experiments were conducted in the atmosphere [1–4]. Faghri [1] qualitatively described the recession process of the liquid layer as well as the likely evaporation/boiling transition with increasing heat flux. Under a low heat flux, heat transfer by conduction and convection prevails. The capillary menisci are formed at the solid-liquid interface. As the heat flux is increased, the menisci recede into the wick due to intensified evaporation. If the heat flux is further increased, nucleate boiling may occur near the heated wall. Li et al. [2, 3] experimentally studied the evaporation/boiling process for a horizontal copper surface sintered with multiple layers of copper mesh. They further indicated that the liquid surface receded non-uniformly at a large heat flux with the lowest point at the center of the heated area. When the liquid thickness around the central area becomes very thin, the effective liquid thin film evaporation is dominant. If the heat flux continues to increase, initial dryout appears at the center of the heated area. Since nucleate boiling is suppressed under thin film evaporation, the ultimate heat transfer limit would be the capillary limit, rather than the boiling limit. However, their experiment was performed at atmosphere pressure and the water level was fixed rather than self-adjusted as in an operating heat pipe. Brautsch and Kew [4] visualized the heat transfer processes of a vertical aluminum heater surface covered with stainless steel meshes capillarily wetted by distilled water. At a low heat flux, small bubbles form first on the heater surface and then in the mesh wick. The bubbles could leave without being trapped by the mesh. As the heat flux is increased, bubbles coalesce into vapor patches which insulate the heater surface from water and lead to intermittent local dryout. Intermittent bursting of vapor patches was observed. The bubbles in the wick and the bursts impeded the liquid return and the heat transfer process.

Wong and Kao [5] performed the first visualization study for operating heat pipes fabricated by a glass container laid with non-sintered copper mesh wicks. They indicated that at low heat loads the heat is dissipated by surface evaporation on the meniscus interface within the mesh wires. Nucleate boiling was observed at increased heat loads. However, under a combination of a fine bottom mesh layer and a fluid charge approximately saturating the wick, nucleate boiling is suppressed at large heat loads. A thin water layer with a large number of menisci is sustained in the fine mesh pores by the strong capillary force therein. Due to the small thermal resistance across the evaporator, the large evaporation area, and the possible contribution of thin film evaporation at the water-wire-vapor interface, both the evaporation temperature and the heating surface temperature are lowest among all the tested wick/charge combinations. It is noted that their mesh wick was *not sintered* and the appearance of nucleate boiling could result from the large thermal resistance across the wick layer.

The performance of a heat pipe is dictated by its maximum heat load, Q_{max}, and thermal resistance, R_{th}. For the capillary-limited maximum heat load, Chi [6] correlated Q_{max} with the figure of merit N consisting of a number of liquid properties of the working fluid. Under a horizontal orientation and with the liquid pressure drop dominating,

$$Q_{max} = 2\left(\frac{\rho_l \lambda \sigma}{\mu_l}\right)\left(\frac{K}{r_{eff}}\right)\left(\frac{A_w}{L_{eff}}\right)\cos\theta, \tag{1}$$

where K is the wick permeability, A_w is the cross-sectional area of the wick, r_{eff} is the effective capillary radius of the wick pores, L_{eff} is the effective length of the liquid flow path, and θ is the apparent contact angle at the evaporating meniscus interlines. Thus, the figure of merit is defined as

$$N \equiv \rho_l \lambda \sigma / \mu_l. \tag{2}$$

For an identical wick structure and heat pipe geometry,

$$Q_{max} \propto N \cos\theta. \tag{3}$$

As for the thermal resistance, the evaporator resistance R_e has been the focus [7–14] since it usually occupies a major portion of the overall resistance of a heat pipe. Kempers et al. [7] compared the heat transfer characteristics of the condenser and the evaporator of a copper–water mesh-wicked heat pipe. With increasing heat load, the condenser resistance varied only slightly, but the evaporator resistance decreased significantly. Such an R_e-Q trend has been widely believed to result from boiling. This argument seems to be supported by the agreement between a boiling heat transfer correlation and the experimental results [7]. On the other hand, Chang et al. [8] argued that R_e decreases with decreasing $\delta_{w,eff}$ in response to increasing heat load. Since the thin-film evaporation at the solid-menisci interlines presents a negligible thermal resistance, R_e mainly comes from the metal path in the wick. Consequently,

$$R_e \approx \delta_{w,eff} / (k_{w,eff} A_e), [K/W] \tag{4}$$

where $\delta_{w,eff}$ is the wetted effective wick thickness, $k_{w,eff}$ is the effective conductivity of the wick, and A_e is the area of the evaporator. Without resorting to boiling, evaporator resistances calculated using Eq. 4 agree with their experimental data showing a trend of decreasing R_e with increasing Q. Hwang et al. [9] also applied Eq. 4 to calculate R_e and obtained a consistent trend with their experiments using reasonably estimated quantities of $\delta_{w,eff}$ and $k_{w,eff}$. Ranjan et al. [10] analyzed the wicking and thin-film evaporation characteristics of microstructures. They suggested that the interline area providing highly effective thin-film evaporation is too small. Another parallel path through the non-thin-film extrinsic menisci should be considered. Their analysis indicated comparable thermal resistances via the thin-film and the non-thin-film paths. This implies that the fluid properties, such as the thermal conductivity, may also be influential to the evaporator resistance. Recently, Bodla et al. [11] made significant progress in analyzing thin-film evaporation from the menisci residing in real sintered-powder wick microstructures. The evaporative heat transfer coefficient, the percentage of thin-film area, and the fraction of mass evaporating from the thin-film region were predicted with respect to apparent contact angle and powder size. Their results also supported the validity of Eq. 4, with negligible thermal resistance of the evaporation from the menisci.

As far as wick capillarity is concerned, its effect is practically important. During the manufacturing process of heat pipes or vapor chambers, a finite exposure time in air is inevitable. After the copper objects are taken out of the sintering furnace, hydrophobic Cu_2O would form on the copper surface with time, faster under a higher temperature. If the object temperature is still high or the elapsed time is too long, the surface wettability, and hence the wick capillarity, would significantly degrade and thereby impair the performance of heat pipes.

In the past few years, our lab has successfully visualized the evaporation characteristics of operating flat-plate heat pipes [12–15]. Combined visualization and evaporator resistance measurements have been conducted for various wicks and working fluids. This article will summarize these achievements and provide insights of the evaporation characteristics in the wicks of heat pipes.

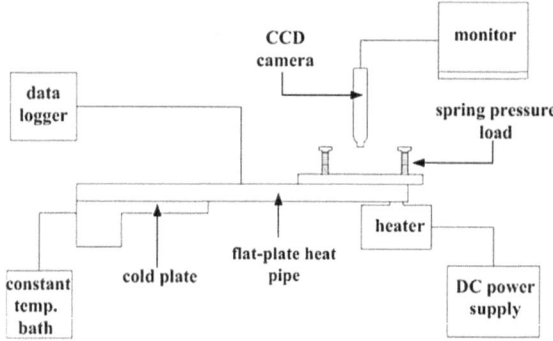

Fig. 1 Overall experimental setup

2 Experimental Methods

A detailed description of the heat pipe structure and experimental procedure is available in Ref. [12]. Here is given a brief description. Figure 1 shows the overall test setup. The 1.1 cm × 1.1 cm heated surface of the heat pipe is connected to the uniform-temperature top surface (1.0 cm × 1.0 cm) of a heating post via a layer of thermal grease. The heating post is carefully insulated except at its top end. The bottom of the other end of the heat pipe is cooled by a 3.0 cm × 3.6 cm cold plate with running water at 17–20 °C. Except for these heated and cooled regions, the heat pipe is well insulated during operation. To prevent condensation on the visualization window, only for recording is the insulation material over the window removed shortly. A CCD camera equipped with a microscopic lens (Optem, Zoom 125) shoots vertically downward through the observation window. High magnification of the evaporation zone can be obtained. Illumination is provided by several high-intensity LED lights.

Figure 2 shows the flat-plate heat pipe designed for visualization and thermal resistance measurement, along with the positions of sixteen implanted thermocouples. These thermocouples, K-type 1 mm-diameter stainless-steel sheathed probe (Omega, Inc.), are denoted by T1–T16, respectively. Thermocouple signals are recorded by a data logger (Fluke, Hydra Series II) with a resolution of 0.1 °C. T1 is measured in the middle of the copper plate at the center of the heated area. Measuring about 1 mm above the center of the evaporator, T2 reflects the vapor temperature leaving the evaporator. T3 represents the vapor temperature at the adiabatic section of the heat pipe. T4 and T5 are the copper plate temperatures in the cooled area. T6 and T7 are the inlet and outlet temperature of the cooling water, respectively. T8–T10 are used to calculate the total heat load Q_t through the copper heating post. T11–T16 are managed to evaluate the lateral conduction through the copper plate, Q_{cond}. Since the copper plate is highly conductive, a cut-off trench is made around the heated area (Fig. 3). With the plate thinned to 1 mm by the trench, lateral conduction is reduced. Furthermore, Q_{cond} is measured using

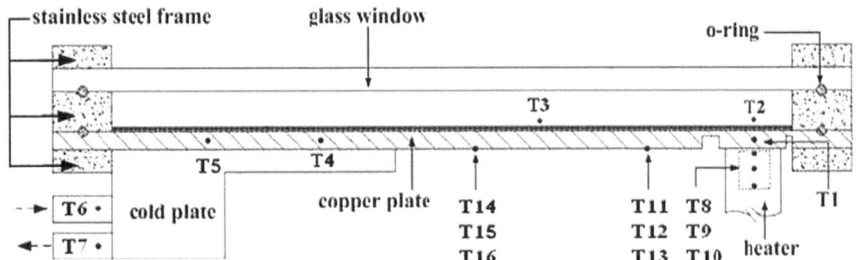

Fig. 2 The flat-plate heat pipe along with thermocouple positions

three pairs of thermocouples (T11–T16) as arranged in Fig. 3b. It is found that the values of the three temperature differences vary by only about 0.1 K. With the average of the three temperature differences, Q_{cond} can be evaluated.

A number of different wicks are adopted for water tests. First are sintered multi-layer mesh wicks. Second are sintered composite powder-mesh wicks with a powdered wick at the evaporator and a mesh wick elsewhere. Either irregular or spherical-shaped powders are used. Third is a sintered homogeneous powdered wick made of irregular-shaped powders. The 100 mesh screen has a wire diameter of 0.114 mm and a wire spacing of 0.14 mm; the 200 mesh screen has a wire diameter of 0.055 mm and a wire spacing of 0.07 mm. Table 1 lists the composition, thickness, porosity, permeability, effective conductivity, and saturate charge for studied sintered-mesh wicks. The wicks are named by the combination of number of layers and the mesh number of the constituent screens, with sequence from top to bottom. For example, a $100 + 2 \times 200$ mesh wick represents that having one 100 mesh top layer and two 200 mesh lower layers. Table 2 lists some important properties of the sintered-powder wicks. For methanol and acetone tests, only the $100 + 200$ mesh wick is used. The wick porosity is determined using the density method. The effective conductivities of the dry wicks are measured following the ASTM D5470 standard test method. Detailed description of our effective conductivity measurement has been given in our earlier work [12]. The uncertainty of them is 5 %.

All the wicks are sintered on a 3 mm thick C1020 oxygenless copper base plate. The mesh wicks are sintered under a fixed pressure in a 850 or 900 °C hydrogen/nitrogen atmosphere for 2 h. The homogeneous powder wicks are loosely sintered on the copper plate under otherwise the same sintering conditions. The composite wicks are sintered in two steps. Two layers of 100 mesh woven screens with a 1.1 cm× 1.1 cm cut-off opening at the heated area are first sintered on the copper base plate under a fixed pressure for 2 h. Then, the copper powders are filled in the opening and loosely sintered for 2 h. Before sintering, the screens and the copper plate are carefully cleaned. The heat pipe has a top glass window for observation. The internal space of the heat pipe is $120 \times 30 \times 7$ mm^3. With the contact surfaces between different pieces sealed with o-rings, the whole structure, including the wicked copper base plate, the top glass window, and the stainless-steel frames, is tightened with bolts. For methanol and acetone tests, anti-corrosion Aflas®

Fig. 3 a Lateral heat
conduction and
b thermocouple arrangement
for its measurement

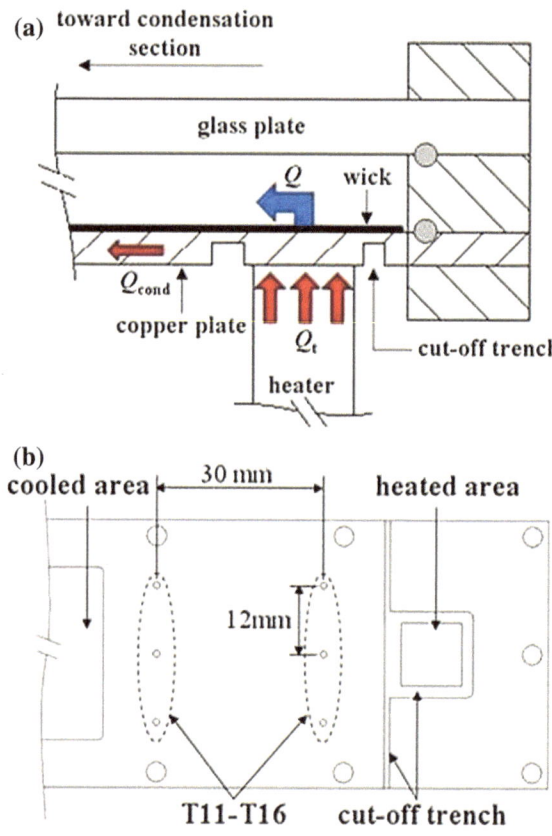

Table 1 Characteristics of different mesh wicks

Wick composition	Thickness (mm)	Porosity[a]	Permeability[b] (m^2)	k^c_{eff} (W/mK)	Saturate charge (ml)
100 + 200 mesh	0.26	0.60	–	13	0.50
2 × 100 mesh	0.32	0.59	1.2×10^{-10}	16	0.56
100 + 2 × 200 mesh	0.34	0.64	–	19	0.65
4 × 200 mesh	0.33	0.59	6.2×10^{-11}	–	0.65

[a] measured using the density method
[b] calculated [1]
[c] measured following ASTM D 5470

o-rings are used. A selected amount of degassed working fluid is filled into the heat pipe right after it has been evacuated down to a pressure of 1×10^{-2} Torr. Such a vacuum degree has been proved to be sufficient in another work [16] in that uniform temperature prevails in the vapor phase without noticeable effect of non-condensable gases. The unavoidable trace amount of non-condensable gases would be pushed to the condenser end by the vapor flow. With our focus on the

Table 2 Characteristics of different sintered-powder wicks

Powder type	Particle size (μm)	Porosity[a]	k_{eff}^{b} (W/mK)
Full irregular (I)	<210	0.62	16
Fine spherical (S_f)	<75	0.37	33
Coarse spherical (S_c)	75 ~ 180	0.47	25

[a] measured using the density method
[b] effective conductivity measured following ASTM D 5470
[c] permeability measured presently

evaporation characteristics, the effects of non-condensable gases can be negligible. Since some working fluid would accumulate at the corners and crevices in the present flat-plate heat pipes, the charges are somewhat larger than the saturate value that exactly fills the wick.

The heat load is increased stepwise from a low heat load, and data are taken under a thermally stable condition for each heat load.

It is noted that the capillarity of the sintered wick degrades in air with time due to surface oxidation. In this work, the copper surface wettability is quantified by the contact angle of a sessile *water* drop on a smooth surface. The contact angle is measured on another plate, which has been taken out of the sintering oven simultaneously with the wicked copper plate, when the heat pipe is evacuated to 1 Torr. Except for the tests on surface wettability effects, the static contact angles on a smooth copper surface are controlled at 10°–12°. When needed, different contact angles are obtained with different elapsed times in the room air. At all these surface conditions, methanol and acetone wet the smooth copper surface.

The evaporator resistance R_e is determined as

$$R_e = (T1 - T2)/Q, \, [K/W] \tag{5}$$

where $Q = Q_t - Q_{cond}$. The net heat flux $q = Q/A_e$, with the heated area $A_e = 1.21 \text{ cm}^2$. In this study, the ratios of Q_{cond}/Q_t prior to local dryout are about 1.1–6 % for water, 5–10 % for both methanol and acetone, smaller for a smaller evaporator resistance.

The superheat at the wick bottom is approximately determined as

$$\Delta T_{sup} \equiv T_w - T_{lv} \approx T1 - T2 - 0.0375q, \tag{6}$$

where the measured vapor temperature T2 closely represents the temperature T_{lv} at the liquid–vapor interface. The 1.5 mm distance from the location of thermocouple T1 to the wick bottom is considered in determining T_w, the temperature at the wall-wick interface.

The uncertainty in the measurements of Q was 7 %. This is determined by the maximum uncertainty associated with the nonlinearity of the temperature readings T8–T10 and the comparison with the output of the DC power supply. For water, the uncertainty in R_e values is also 7 %, since the uncertainties primarily arise from those of Q, according to the definition of R_e in Eq. 5. For methanol and acetone, the 0.1 K thermocouple resolution limit corresponds to a maximum

Table 3 Thermophysical properties of water, methanol and acetone at selected temperatures

	ρ_l (kg/m^3)	λ (kJ/kg)	σ (N/m)	μ_l (cP)	k_l (W/mK)	N(W/m^2)^a
Water^b	996	2406.7	0.0695	0.799	0.632	2.09×10^{11}
Methanol^c	791	1165	0.0227	0.611	0.204	3.42×10^{10}
Acetone^c	790	552	0.0237	0.323	0.181	3.20×10^{10}

^{a} figure of merit $N \equiv \rho_l \lambda \sigma / \mu_l$
^{b} at 40 °C, but ρ_l and μ_l at 30 °C
^{c} at 20 °C

uncertainty of 10 % in ΔT_{sup} at the lowest heat loads (cf. Fig. 30). This results in maximum uncertainties of 12 % in R_e for methanol and acetone.

The figures of merit of the three different working fluids, as well as the related fluid properties, at selected temperatures are shown in Table 3. The temperatures for ρ_l and μ_l roughly correspond to the average of the evaporator and condenser temperatures at q_{max}, while those for λ, σ and k_l roughly correspond to the evaporation temperatures approximated by T2 (cf. Fig. 25).

3 Results and Discussion

3.1 Water in Mesh and Powder Wicks

3.1.1 Water in Multi-Layer Mesh Wicks

Results for two-layer mesh wicks are first discussed because the liquid level within the evaporator can be observed down to dryout at the bottom. In this work, observation is positioned at the center or the front end (5 mm upstream from the center) of the heated area. For a specific wick composition, the heat pipe performance with respect to water charge (v) is first examined. Figure 4 compares the evaporator resistances (R_e) for a $100 + 200$ mesh wick at different charges. The saturate charge to exactly fill this wick is 0.5 ml. Excess charge is made to account for the accumulation at the corners and crevices in the heat pipe. In general, for a specific charge R_e gradually decreases with increasing q until a minimum value $R_{e,min}$. Then, a sharp increase in R_e appears as a result of the occurrence of local dryout. The heat load with $R_{e,min}$ is called the maximum heat load, q_{max}. The $R_{e,min}$ for the largest charge ($v = 0.92$ ml) is obtained at a heat flux of $q = 72$ W/cm^2. With a smaller charge of 0.85 ml, the heat flux yielding $R_{e,min}$ shifts to 57 W/cm^2. For the smallest charge of 0.76 ml, the heat flux with the $R_{e,min}$ further shifts to 29 W/cm^2, and local dryout appears at a heat flux of 39 W/cm^2. In the following, detailed discussion will be made for the case with $v = 0.85$ ml.

Figure 5 shows the corresponding visualization at the center and the front end of the heated area. The corresponding evaporator resistances have been given in Fig. 4. At a low heat load, a small part of the upper mesh is exposed (Fig. 5a, $q = 18$ W/cm^2) and the water surface is pretty smooth. The corresponding

0.18
0.16
0.14
0.12
0.1
0.08
0.06
0.04
0.02
0
R_e(K/W)
0
20
40
60
80
100
q (W/cm²)
v=0.92ml
v=0.85ml
v=0.76ml

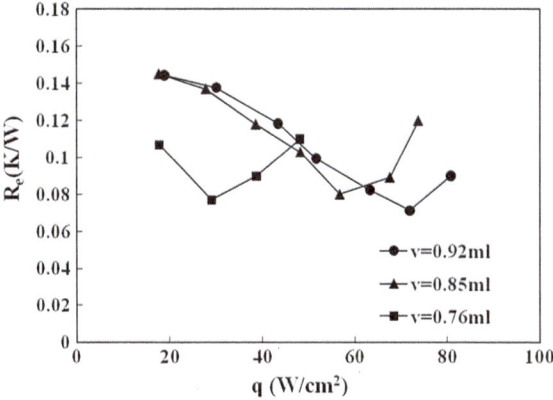

Fig. 4 Evaporator resistances of 100 + 200 mesh wick at various charges and heat fluxes

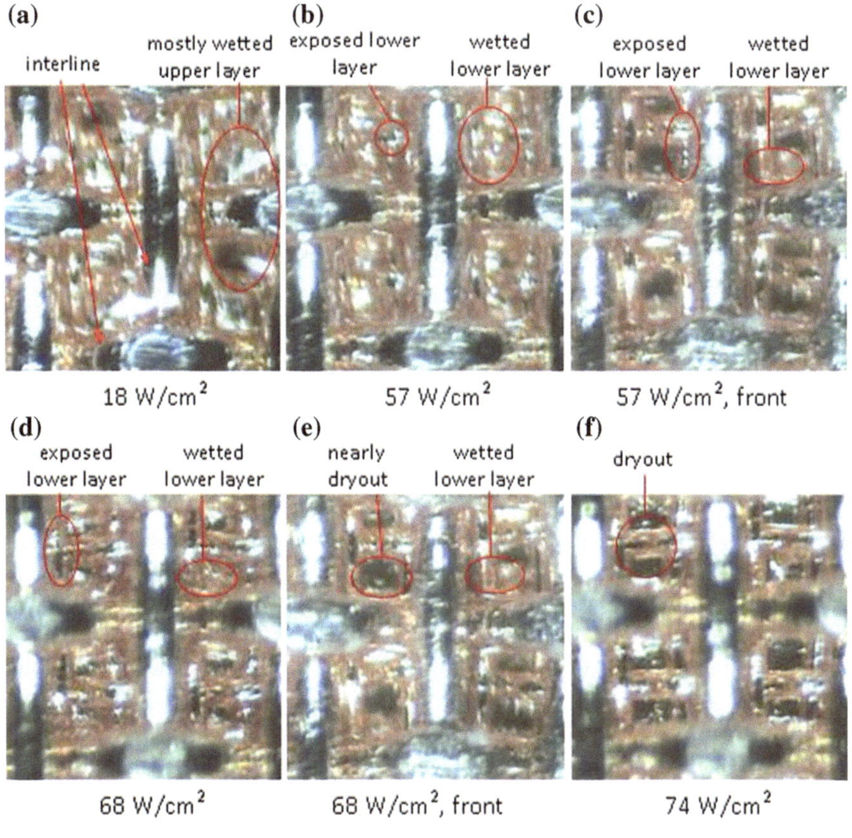

(a)
interline
mostly wetted upper layer
18 W/cm²
(b)
exposed lower layer
wetted lower layer
57 W/cm²
(c)
exposed lower layer
wetted lower layer
57 W/cm², front
(d)
exposed lower layer
wetted lower layer
68 W/cm²
(e)
nearly dryout
wetted lower layer
68 W/cm², front
(f)
dryout
74 W/cm²

Fig. 5 Visualization of evaporation process for 100 + 200 mesh wick at various heat fluxes under $v = 0.85$ ml

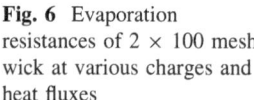

Fig. 6 Evaporation resistances of 2 × 100 mesh wick at various charges and heat fluxes

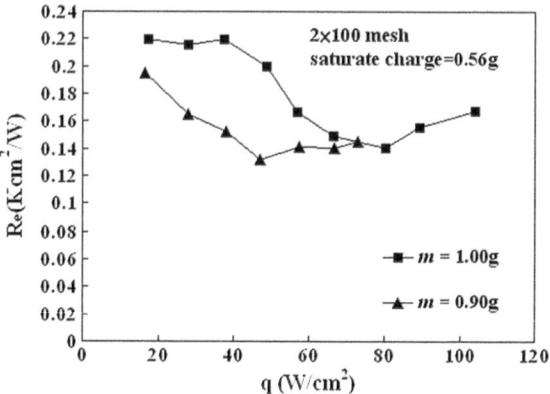

evaporator resistance is relatively large (Fig. 4). With increasing heat load, the water layer recedes and the evaporator resistance reduces concurrently. At $q = 57$ W/cm^2, the water layer recedes to the lower layer with its wire peaks exposed (Fig. 5b). At the front end, the lower layer is more exposed but thin water layer appears to sustain in the mesh hole (Fig. 5c). The $R_{e,min}$ for $v = 0.85$ ml is found for this heat flux (Fig. 4). At $q = 68$ W/cm^2, the water layer at the heated area center further lowers (Fig. 5d). However, Fig. 5e at the front end does not clearly display dryout, although local dryout could have occurred according to the mild increase in R_e. At $q = 74$ W/cm^2, dryout extends to the heated area center, as reflected by the clear and glossy image of bottom mesh wires (Fig. 5f). As a result, R_e increases significantly. Highly distorted images in Fig. 5 imply corrugated meniscus water surfaces.

Figure 6 shows the evaporation resistances at different charges and heat loads for a 2 × 100 mesh wick. Again, the heat flux yielding the minimum R_e is smaller for a smaller charge. But the values of the minimum R_e are larger than for the 100 + 200 mesh wick. Figure 7 shows the corresponding visualization at $v = 0.90$ ml. Similarly, the water level lowers with increasing q, as reflected in Fig. 7a, b, c. At $q = 47$ W/cm^2 the upper part of the lower layer begins to be exposed, and at 67 W/cm^2 more exposure is observed. In Fig. 7b the diameter of the bottom wires appear shrunk under the concave menisci. The minimum R_e is found at this situation, with a thin water film sustained within the bottom layer. Dryout occurs at the front end for $q = 67$ W/cm^2 (Fig. 7d), leading to a slight rise of R_e (Fig. 6).

Experiments were also conducted for a 100 + 2 × 200 mesh wick, which has similar wick thickness as the 2 × 100 mesh wick (Table 1). The lower layers of fine 200 mesh screen provide stronger capillary force, larger meniscus surface area, and possibly more meniscus-wire interfaces with effective thin film evaporation. The evaporation resistances at different charges and heat loads are given in Fig. 8; visualization corresponding to $v = 0.88$ ml is shown in Fig. 9. At $q = 28$ W/cm^2, the middle layer is still immersed (Fig. 9a). At $q = 37$ W/cm^2, the

Fig. 7 Visualization of evaporation process for 2 × 100 mesh wick at various heat fluxes under $v = 0.90$ ml

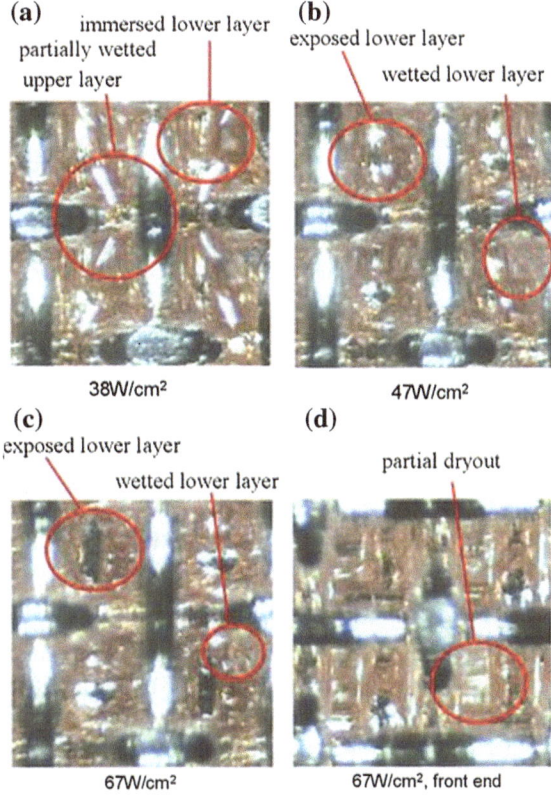

Fig. 8 Evaporation resistances of 100 + 2 × 200 mesh wick at various charges and heat fluxes

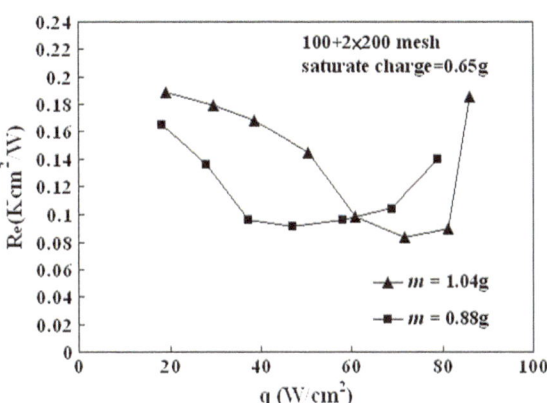

middle layer becomes partially exposed, but the bottom layer is still immersed (Fig. 9b). The recession of water film causes significant reduction in R_e (Fig. 8). With q increased to 47 W/cm², the peaks of the bottom layer appear at the front

Fig. 9 Visualization of evaporation process for $100 + 2 \times 200$ mesh wick at various heat fluxes under $v = 0.88$ ml

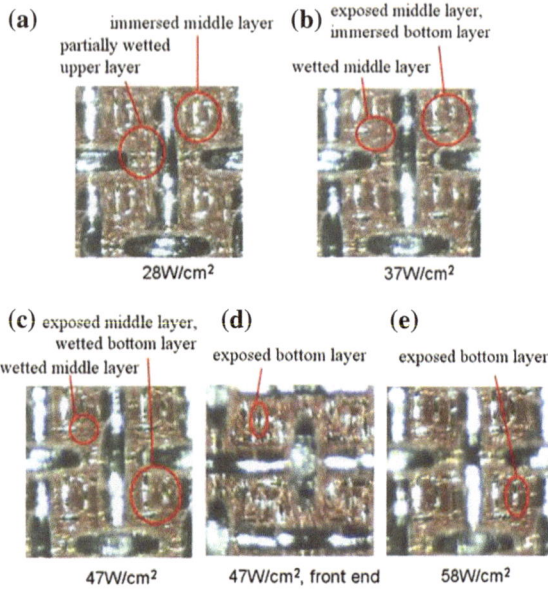

end (Fig. 9d). At this heat flux is obtained the minimum R_e, which is only slightly less than the proximate values (Fig. 8). At $q = 58$ W/cm², the bottom layer peaks at the center also appear. Further water recession and rise of R_e is observed with further increasing heat flux. However, dryout is difficult to discern under the three mesh layers. For the case with a larger charge $v = 1.04$ ml, the values of R_e are larger at low heat fluxes, and the minimum value occurred at a heat flux as large as $q = 72$ W/cm². For both charges, the $100 + 2 \times 200$ mesh wick yields smaller values of minimum R_e than the 2×100 mesh wick did. In addition, these values approximately equal those of the $100 + 200$ mesh wick (Fig. 4). As will be discussed, the capillarity of the bottom mesh layer is important to evaporation performance under large heat fluxes.

Figure 10 presents two representative cases of the superheat measurement. In Fig. 10a are the results for $100 + 200$ mesh wick at various charges. The corresponding evaporation resistances are available in Fig. 4 and the visualization images for $v = 0.85$ ml are shown in Fig. 5. The superheats for the 2×100 mesh wick at various charges are given in Fig. 10b. Their corresponding evaporation resistances and the visualization images for $v = 0.9$ ml are available in Figs. 6 and 7, respectively. It can be seen that at small charges the superheats increase monotonically before local dryout occurs. At a larger charge, the superheats may first increase with increasing heat flux to a maximum and then decrease with film thinning. When dryout occurs, the superheats re-rise significantly. In the case of Fig. 10a and most other test runs, the maximum superheats prior to the occurrence of dryout are less than 5 K. Only in the case of the 2×100 mesh wick with a large charge of $v = 1.0$ ml, shown in Fig. 10b, does the maximum superheat reach 8 K.

Fig. 10 Representative
superheats at the evaporator
for various wicks under
different heat fluxes,
a 100 + 200 mesh wick,
b 2 × 100 mesh wick

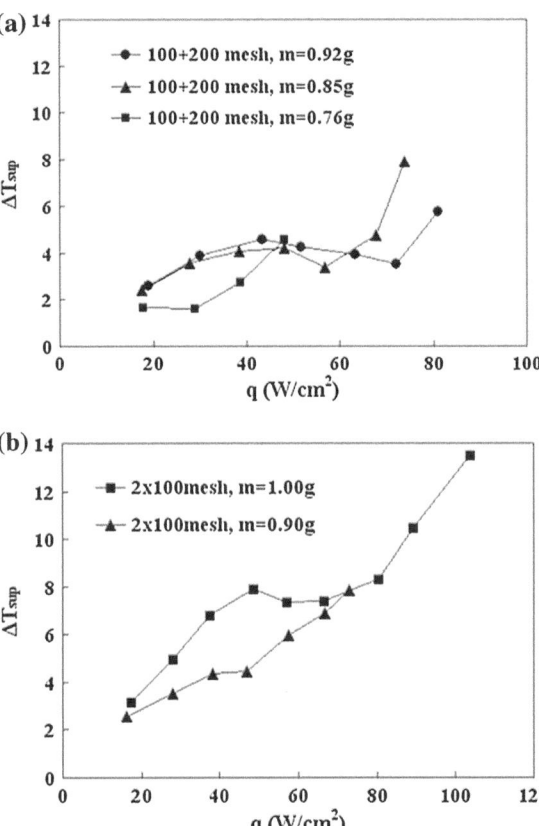

In the work of Wong and Kao [5] for non-sintered two-layer copper wicks, intense nucleate boiling occurs beyond certain threshold heat loads. Presently, however, limited superheat exists at the wick bottom due to much smaller thermal resistances associated with the high conductivity of sintered copper wicks [2]. Consequently, nucleate boiling is absent.

Figure 11 compares the evaporator resistances for different two-layer wicks with selected charges. Also included is an additional set of data for 200 + 100 mesh wick, with the coarse mesh at the bottom. The values of $R_{e,min}$ for the two wicks having a coarse bottom layer are similar, both larger than that of the wick having a fine bottom layer. Wong and Kao [5] also observed similar advantages using a fine bottom mesh. The reasons include: (1) thinner water layer can be sustained at larger heat loads with fine bottom mesh; (2) the more corrugated menisci within the fine mesh provide a larger evaporation area; (3) the more water-wire-vapor interfaces where a very thin capillary layer is sustained to exhibit effective thin film evaporation. For the two coarse-bottom-layer wicks, R_e of the thinner 200 + 100 mesh wick rises sharply after local dryout appears. This is because the liquid flow resistance is larger for a thinner cross-section. With a

Fig. 11 Comparison between evaporator resistances of different two-layer wicks at various heat fluxes

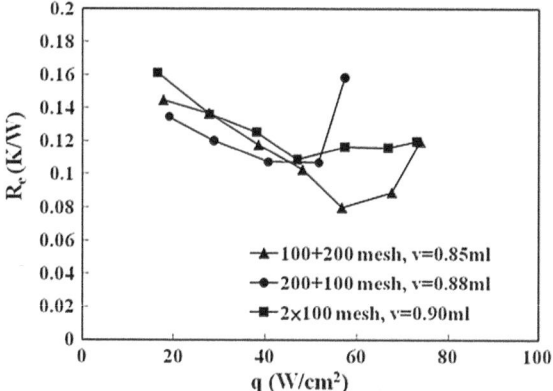

Fig. 12 Comparison between evaporator resistances of different wicks with similar thicknesses at various heat fluxes

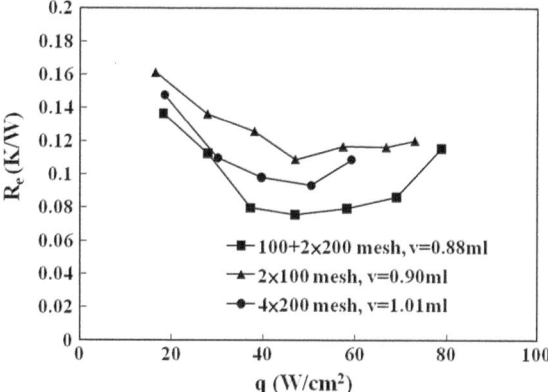

larger flow resistance and weaker capillary force provided by the coarse bottom layer, dryout extends more rapidly.

In Fig. 12 are compared the evaporator resistances for three different wicks having similar thicknesses of about 0.33 ± 0.01 mm. The 2×100 mesh wick has weakest capillarity but highest permeability, while the 4×200 mesh wick has strongest capillarity but lowest permeability. The $100 + 2 \times 200$ mesh wick has a high-capillarity bottom layer as well as high permeability. These characteristics influence their evaporation performances. As shown in Fig. 12, the $100 + 2 \times 200$ mesh wick yields not only a smallest $R_{e,min}$, but also a wide range of heat flux with small values of R_e. The $R_{e,min}$ of the 2×100 mesh wick is the largest. However, its higher permeability prevents a sharp rise of R_e for a relatively wide range of heat flux after the occurrence of local dryout. In comparison, $R_{e,min}$ of the 4×200 mesh can not be as small as that of the $100 + 2 \times 200$ mesh wick, although both have a fine bottom mesh layer. Furthermore, R_e rises sharply after local dryout occurs, even though the charge is relatively large. Such behavior can be attributed to its low permeability that retarded the water flow.

Fig. 13 Evaporator resistances for the composite irregular-powder wick at various charges and heat fluxes

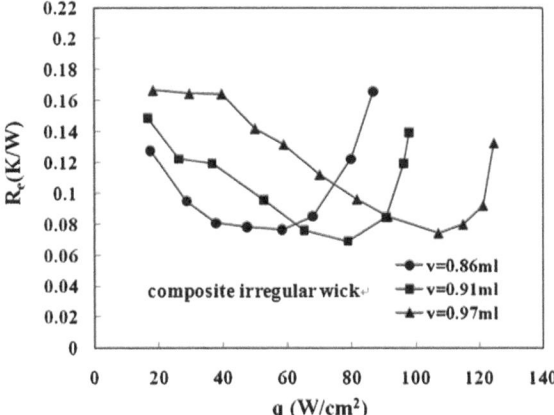

No nucleate boiling is observed in all the above tests. Quiescent surface evaporation prevails. To investigate whether nucleate boiling would occur for a thicker mesh wick, we examined for a $4 \times 100 + 2 \times 200$ mesh wick with thickness of about 0.8 mm. With a large charge for this thick wick, the values of R_e are larger at low heat loads, due to the thick water layer in the wick. In addition, $R_{e,min}$ and local dryout occur at a large heat flux of 150 W/cm^2, because of the larger charge and higher wick permeability. Still, no nucleate boiling is observed up to $q = 160$ W/cm^2.

3.1.2 Water in Sintered-Powder Wicks

The sintered-powdered evaporator wicks were tested in two wick arrangements. The first was a composite wick with powders only at the evaporator and 2×100 mesh wick elsewhere. This arrangement was used to separate the effects of the high capillarity and low permeability of the powder wicks. In the second arrangement of homogeneous powder wicks, the integral effects of high capillarity and low permeability were examined.

Composite Wicks with Irregular-Powder Evaporator

Experiments were first conducted for the widely-dispersed irregular-powder wick, coded as I in Table 2. In this study, a uniform thickness of 0.32 mm was maintained for all the composite powder-mesh wicks. Figure 13 shows the evaporator resistances versus heat flux for three water charges. The smallest charge $v = 0.86$ ml yields smallest evaporator resistances at low heat fluxes. The evaporator resistance rapidly decreases to 0.081–0.077 K/W at 38–58 W/cm^2. Similar to the mesh wicks, q_{max} is higher for a larger charge and the values of $R_{e,min}$ are

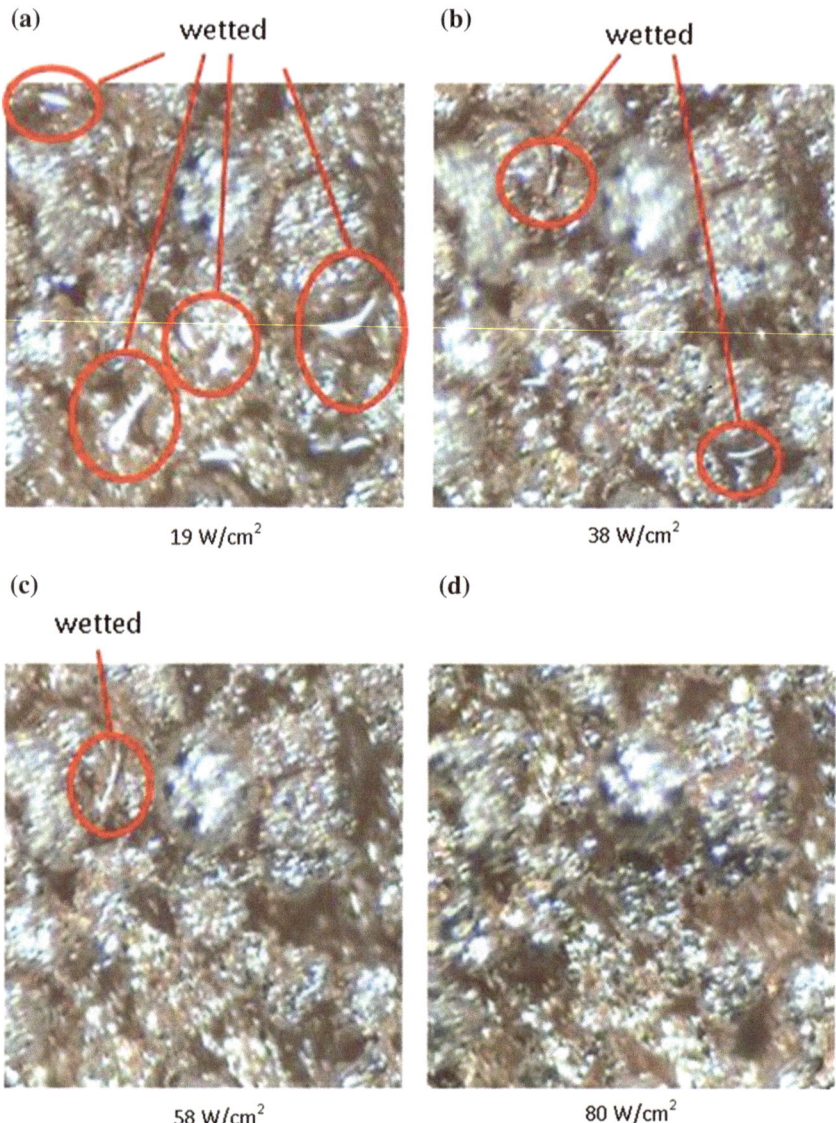

Fig. 14 Visualization of the evaporation process for the composite irregular-powder wick at various heat fluxes under $v = 0.86$ ml

similar. Figure 14 compares the water levels under $q = 19$–83 W/cm^2 for $v = 0.86$ ml. These photographs were taken at the center of the evaporation area. At $q = 19$ W/cm^2, the wick is mostly emerged in water (Fig. 14a). At $q = 38$ W/cm^2, powders on the top layer are all exposed and water can only be observed at certain depth in the wick (Fig. 14b). The water layer further recedes at $q = 58$ W/

Fig. 15 Schematic illustration of the correlation between evaporator resistance and water film recession

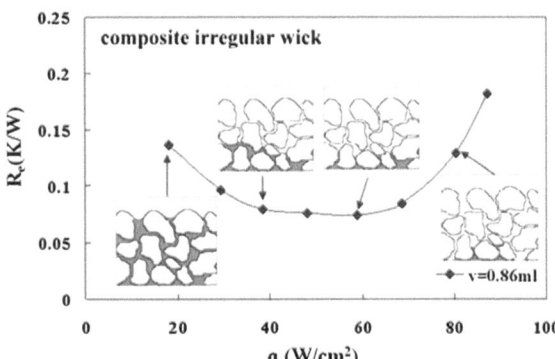

cm^2 (Fig. 14c). At $q = 80$ W/cm^2, when dry-out prevails (Fig. 13), water is no longer observable (Fig. 14d). Similar to the multi-layer mesh wicks, the smaller values of R_e at $q = 38–58$ W/cm^2 result from the smaller thermal resistances associated with the thin water layers sustained at the wick bottom.

Figure 15 illustrates the variation of the evaporator resistance with the water layer recession according to the data for the case of $v = 0.86$ ml in Fig. 13 and the corresponding images in Fig. 14.

No nucleate boiling is observed in these tests, with quiescent surface evaporation prevailing. This is similar to the sintered multi-layer mesh wicks, even though there are a large amount of nucleation sites in the irregular-powder wicks. Here in the sintered-powder wicks, superheat is still insufficient to activate nucleate boiling.

The variations of the superheat with increasing heat flux for the above different charges are illustrated in Fig. 16. Generally, the superheat increases with increasing q at low values of q. Prior to the occurrence of local dryout, the maximum superheats range from 3 to7 K for these three cases. However, for the two cases with larger charges, the superheat begins to decrease with increasing q after the maxima, in response to the decrease in the liquid layer thickness.

When local dryout appears, the superheat rises again. For the case with the smallest charge, the superheat increases monotonically. This is because the liquid layer is already thin at low heat fluxes and local dryout occurs earlier to lead to monotonic increase of the superheat.

Composite Wicks with Spherical-Powder Evaporator

Experiments were also conducted for the sintered wicks having spherical powders. To reveal the effects of powder size, coarse (S_c) and fine (S_f) powders were investigated. The properties of these sintered spherical powders are given in Table 2. It is noted that since the effective conductivity and porosity of the three powder wicks are different, effects of effective conductivity or porosity can also be

Fig. 16 Superheats versus
heat flux for various charges
for the composite irregular-
powder wick

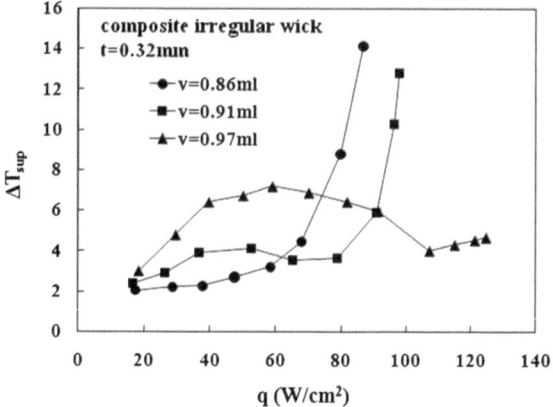

Fig. 17 Evaporator
resistances for different
spherical-powder wicks at a
similar charge of
$v = 0.90 \pm 0.01$ ml

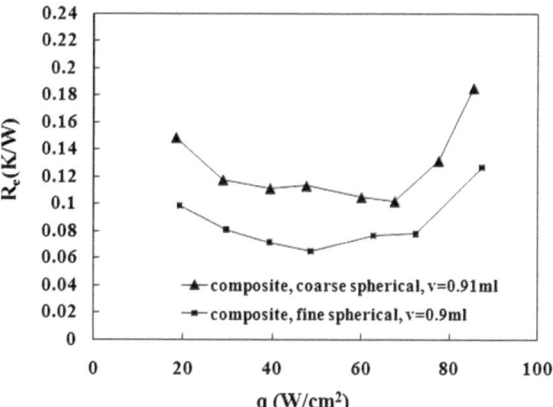

revealed. Figure 17 compares the evaporator resistances associated with different spherical-powder wicks at a charge of $v = 0.90 \pm 0.01$ ml.

Figure 18 shows the corresponding images of both wick evaporators at low heat fluxes. In Fig. 18, wick S_c exhibits a larger $R_{e,min}$ of 0.107 K/W than the 0.066 K/W of wick S_f. However, the q_{max}s of the two wicks are not significantly different. The absence of fine powders lead to not only weaker capillarity but also smaller effective conductivity (Table 2) of wick S_c. The highly capillarity fine-powder wick S_f exhibits better evaporation performances with smaller $R_{e,min}$s. In addition, wick S_f exhibits smallest evaporator resistances at small heat fluxes. This results mainly from its high effective conductivity (Table 2). Since the wick is densely packed, as can be seen in Fig. 18a and from its low porosity given in Table 2, the heat conduction mode dominates in such a thick-liquid-layer situation. Although the thin-film area and the evaporation rate are larger for the fine-powder wick [11], this provides limited contribution to the improvement of evaporator resistance, which is dominated by the thermal path in the wick (Eq. 4). Near the q_{max}, the water layer

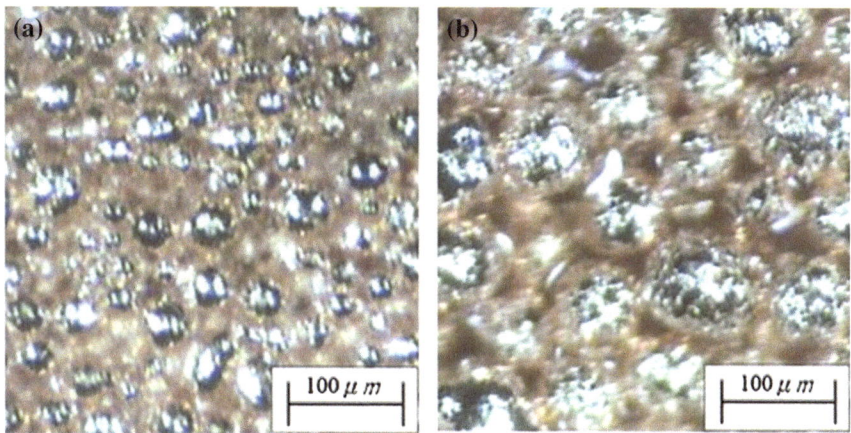

fine spherical-powder wick   coarse spherical-powder wick

Fig. 18 Visualization of the evaporator for the composite spherical-powder wicks at low heat fluxes, **a** fine spherical-powder wick, $v = 0.90$ ml, $q = 19$ W/cm^2; **b** coarse spherical-powder wick, $v = 0.91$ ml, $q = 18$ W/cm^2

deep within the wick becomes indiscernible because of blockage by the powders. Furthermore, the morphology of spherical powders makes it difficult to distinguish the reflective light from the water layer surfaces and from the powder surfaces. Nonetheless, it can be inferred that thinner water layer could be sustained at the wick bottom with the presence of fine powders.

Homogeneous Irregular-Powder Wicks

To further identify the effect of wick permeability, we measured the evaporator resistances for homogeneous irregular-powder wick *I* with thicknesses of 0.4 mm and 1.0 mm. For a 0.4 mm thickness, the low permeability of the sintered-powder wick obviously retards the condensed water from returning to the evaporator. Generally, more water is detained at the condenser than in the composite wick tests, even when the water level in the evaporator is low.

This effect exerts more influence in low-charge cases. In Fig. 19, the cases of $v = 0.93$ and 0.97 ml (the saturate charge was 0.74 ml) exhibit smaller q_{max}s at low heat fluxes, in comparison with the counterparts of the powder-mesh composite wick shown in Fig. 13. The early occurrence of local dryout leads to large $R_{e,min}$ values of 0.10 K/W. With a larger charge of $v = 1.06$ ml, the local dryout can be postponed and the $R_{e,min}$ reduces to 0.074 K/W. Figure 20 shows the evaporator resistances for 1.0 mm-thick homogeneous irregular-powder wick at various charges. The saturate charge for this wick is 1.85 ml. When $v = 1.5$ ml, the initial R_e is small but local dryout occurs early at 60 W/cm^2. Even at this time, water can

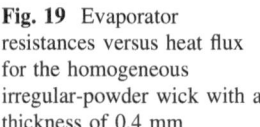

Fig. 19 Evaporator resistances versus heat flux for the homogeneous irregular-powder wick with a thickness of 0.4 mm

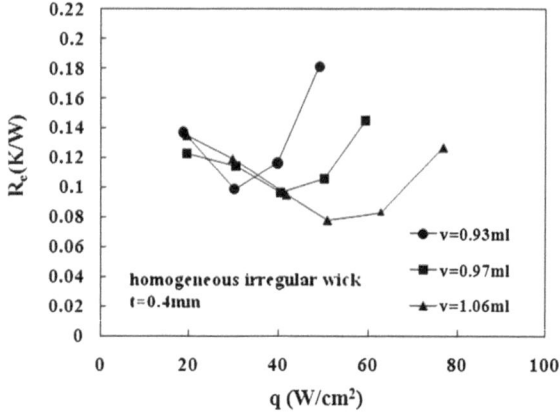

be seen to accumulate at the condenser. At $v = 1.8$ ml, the minimum R_e can reach 0.074 K/W. This value is similar to the minimum values of 0.066–0.074 K/W for the composite wicks containing fine pores at the bottom (Figs. 13 and 17). The q_{max} is about 130 W/cm^2. At a larger charge of $v = 2.0$ ml, the applied heat flux can reach 170 W/cm^2 without local dryout. By increasing the wick thickness, the liquid flow resistance is reduced so that the strong capillarity of irregular powders is not overshadowed. Consequently, in applying a sintered-powder wick in heat pipes, its low permeability needs to be compensated with either a larger wick thickness and/ or a larger charge, provided that similar heat load limits and minimum evaporator resistances as for sintered mesh wicks are to be achieved.

The superheats versus q for the homogeneous irregular wick of $t = 1.0$ mm are shown in Fig. 21. The corresponding R_e variations are given in Fig. 20. The trend in superheat variation is similar to that for the composite irregular wick shown in Fig. 16. In the present cases, a maximum value of 12 °C is obtained without nucleation.

Nucleate boiling is not observed in all the above test conditions for the homogeneous irregular-powder wicks. In powder wicks, the water layer can no longer be observed after the heat flux approaches the critical value. But in these cases there is no sign of water eruption from the wick bottom. Observations have benn further made for more likely conditions, i.e., a thicker wick, a larger charge, higher heat fluxes, and elevated cooling water temperatures. For the 1 mm-thick full irregular-powder wick at a charge of $v = 2.0$ ml (cf. Fig. 20), heat fluxes up to 148 W/cm^2 are applied at a cooling water temperature of 35 °C. In this specific case, the vapor temperature above the evaporator (T2) is 64.1 °C and the measured superheat is as large as 13.7 °C. With the cooling water temperature further raised to 45 °C, we tested up to a heat flux of 117 W/cm^2. No more higher flux was made due to the excessive heater temperature. Still, no nucleate boiling was observed for these conditions. In comparison, we observed that a superheat less than 1 K is sufficient to activate water boiling for a wicked surface when the filling tube of the heat pipe is open to the atmosphere.

Fig. 20 Evaporator resistances versus heat flux for the homogeneous irregular-powder wick with a thickness of 1.0 mm

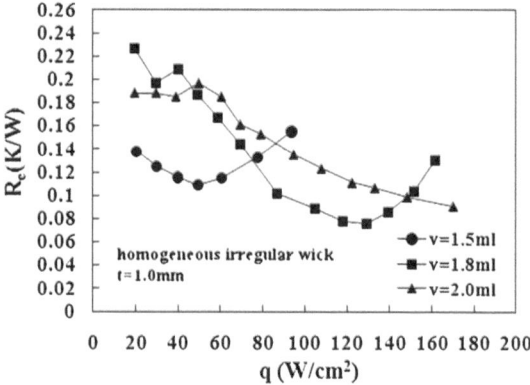

Fig. 21 Superheats versus heat flux for various charges for the homogeneous irregular-powder wick with $t = 1.0$ mm

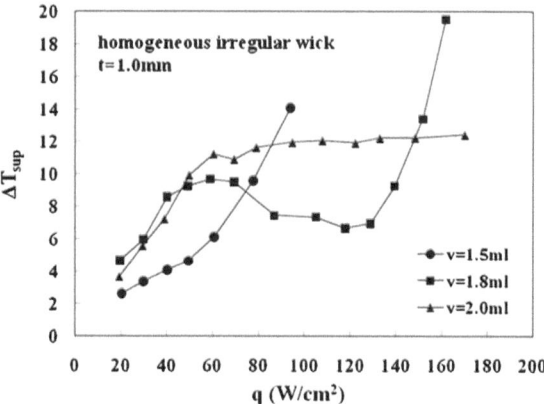

The fact that nucleate boiling of water is absent with significant superheats in sintered mesh or powder wicks of operating heat pipes is probably due to the different situation above the evaporating surface. In operating heat pipes, the vapor from the evaporating surface rapidly rushes to the condenser due to a large pressure difference. The vapor concentration fails to become saturated above the surface. With the high-energy molecules escaping from the liquid surface more easily, larger superheat threshold is required for nucleate boiling. However, this speculation calls for further examination.

Since nucleate boiling is absent in heat pipes with water and sintered copper-mesh or copper-powder wicks thinner than 1 mm, the most common heat pipes for electronics cooling, the amount of nucleation sites in selecting a certain wick is unimportant as far as the thermal performance is concerned. However, the elevated superheats for non-sintered wicks [5], thicker sintered wicks [17] or immersed wicks due to overcharge or gravity still may induce nucleate boiling under high heat fluxes. In addition, nucleate boiling also may occur for heat pipes adopting working fluids having smaller surface tensions, as will be discussed later.

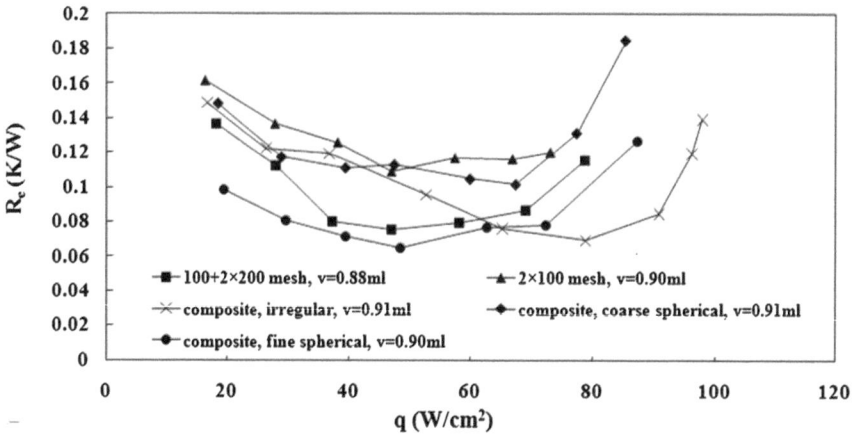

Fig. 22 Comparison of the evaporator resistances between the composite powder-mesh wicks and mesh wicks at similar wick thickness and water charge

3.1.3 Comparison Between Powder and Mesh Evaporators

Comparison is first made between multi-layer mesh wicks and composite powder-mesh wicks. Such comparison focuses on the effects of wick capillarity by minimizing the effects of wick permeability. Figure 22 compares the evaporator resistances versus heat flux for different wicks at similar wick thickness $(0.33 \pm 0.01$ mm) and water charge $(v = 0.90 \pm 0.02$ ml). Large values of $R_{e,\min} = 0.11$–0.12 K/W, are obtained for both the 2×100 mesh wick and the composite wick comprising only coarse powders (S_c). For the composite wicks containing fine powders, S_f and I, and the homogeneous $100 + 2 \times 200$ mesh wick, the minimum values of R_e can reach down to 0.066–0.074 K/W. This is because thin water layers can be sustained under high heat fluxes in wicks having highly capillary fine pores at the wick bottom.

Comparison is further made between multi-layer-mesh wicks and homogeneous sintered-powder wicks, with the effects of wick permeability also involved. Figure 23 shows that at similar wick thicknesses $(0.32$–0.4 mm) and similar charges $(1.04$–1.06 ml), the lower permeability of the sintered-powder wick results in earlier dryout than the $100 + 2 \times 200$ mesh wick did.

3.2 Different Working Fluids in 100 + 200 Mesh Wick

It is noted that the sintering temperature for the wicks in the following experimental data was 850 °C, while it was 900 °C for the data presented earlier. Different sintering temperatures do affect the evaporation performance. But such effects are beyond the scope of the present work. Figure 24 compares the

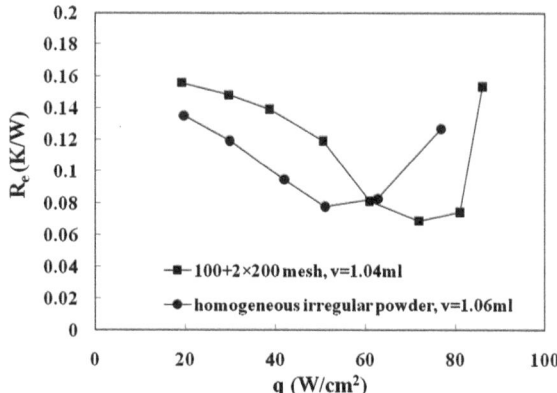

Fig. 23 Comparison of the evaporator resistances between the homogeneous powder wicks and mesh wicks at similar wick thickness and water charge

evaporator resistances versus heat flux under different charges for three working fluids: water, methanol and acetone. Water exhibits much larger q_{max}s than the others. The case of $v = 1.0$ ml for water is not included because dryout does not occur up to a heat flux of 120 W/cm². No higher heat flux was made due to the excessive heater temperature. It is noted that the performance may vary slightly in the repeated tests. For example, the q_{max}s can be only about 90 W/cm² in the repeated tests other than the case of $v = 0.9$ ml for water shown in Fig. 24a. Methanol exhibits much smaller q_{max}s than water, and acetone exhibits the smallest values of about 13–25 W/cm². According to the visualization, very weak nucleate boiling was sometimes observed for methanol near q_{max} for all the fluid charges. In typical cases, the nucleate boiling was only observed near the center of the evaporator, and the activity was too weak to agitate the working fluid. It is noted that nucleate boiling was not observed in every test for methanol. Weak nucleation was observed for acetone near q_{max} for all the three charges. The boiling intensity is stronger than for methanol. Slow periodic up-and-down motion of the liquid surface results from the cyclic bubble growth, coalescence and collapse under the mesh. Nucleate boiling could be observed over a significant portion of the evaporator. The weak nucleation for acetone could be detected by the thermocouple implanted under the evaporator in slow fluctuations of about ±0.5 K. For methanol, the nucleation intensity was too weak to be detected. For most of the cases, nucleate boiling was suppressed beyond q_{max}. The weak and slow bubbling process did not likely block the capillary flow. Consequently, the heat transfer in the present methanol and acetone tests should still be capillary-limited. More discussion about the behavior and effects of the nucleate boiling will be made later on together with the superheat data.

Figure 25 presents the temperature measurements of T1 and T2 versus heat flux for typical tests at $v = 0.9$ ml. These temperatures are at 18–28 °C for methanol and acetone before local dryout. For water, values of T1 and T2 grow with the heat flux up to 50 °C and 40 °C, respectively, at $q_{max} = 104$ W/cm².

Fig. 24 Evaporator
resistances versus heat flux
for various working fluids at
different charges, **a** water,
b methanol, and **c** acetone

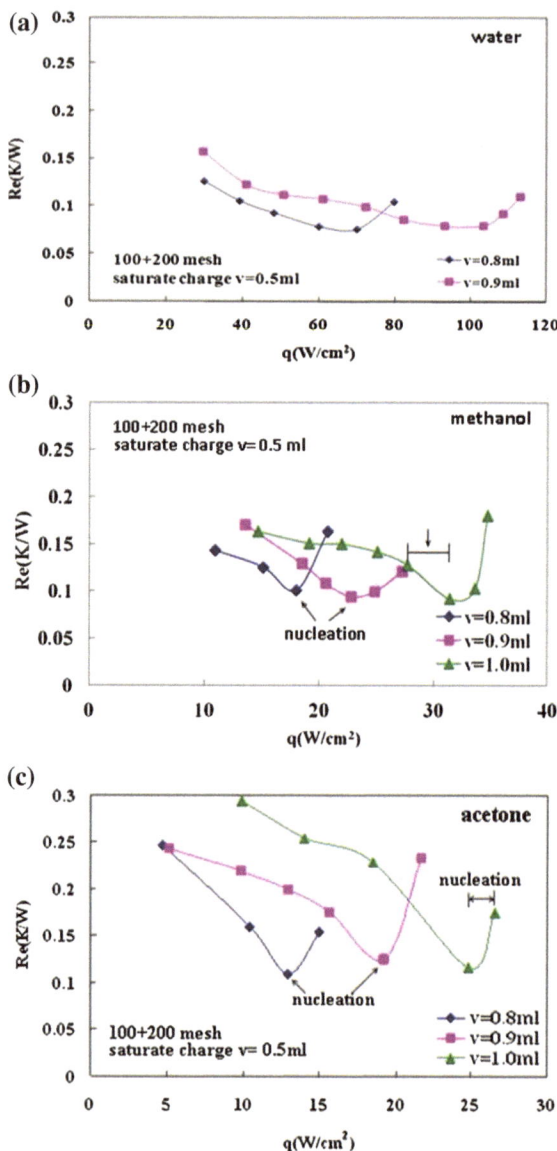

In Fig. 26, the evaporator resistances for different working fluids are compared
for $v = 0.9$ and 0.8 ml, respectively. It is known that the Q_{max} for a working fluid
can be related with its figure of merit, N [6], as in Eq. 3. Although the static
contact angles are zero for methanol and acetone and $10°–12°$ for water, amplified
apparent angles would be present during heat pipe operation [18]. The apparent
contact angles can not be measured, but the values of θ for water should be larger
than those of methanol and acetone. Therefore, the Q_{max}s of the three working

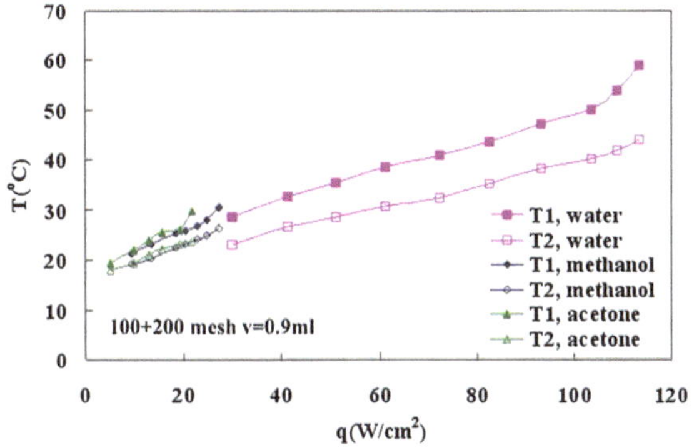

Fig. 25 Plate temperature under the evaporator (T1) and vapor temperature above the evaporator (T2) versus heat flux for water, methanol, and acetone at the same charge of $v = 0.9$ ml

Fig. 26 Evaporator resistances versus heat flux for water, methanol, and acetone at the same charges, **a** $v = 0.9$ ml and **b** $v = 0.8$ ml

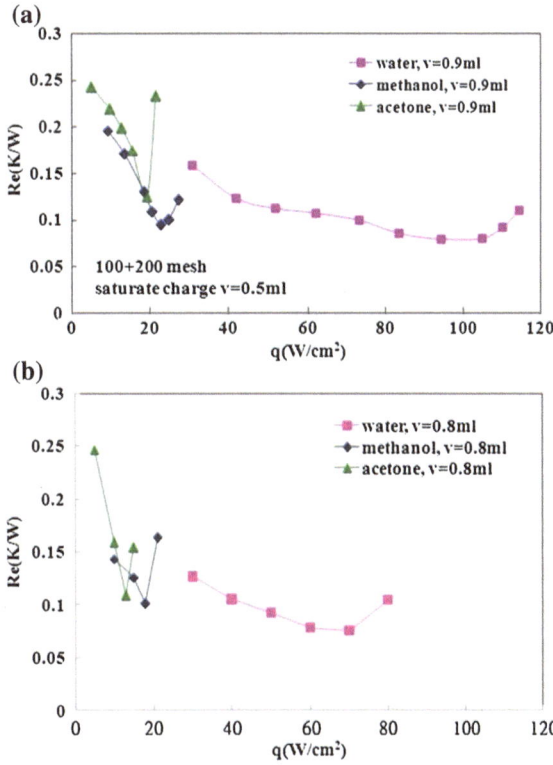

fluids would be primarily proportional to their figures of merit but modified by the differences in $\cos\theta$. In Fig. 26 the Q_{max}s of different fluids are compared under two fixed fluid charges. Table 4 further compares the ratios between various q_{max}s against the ratios of figures of merit. While $N_w/N_m = 6.1$, $N_w/N_a = 6.5$, and $N_m/N_a = 1.1$, the ratios of q_{max} are $q_{max,w}/q_{max,m} \sim 3.9$–$4.6$, $q_{max,w}/q_{max,a} \sim 5.4$ and $q_{max,m}/q_{max,a} \sim 1.2$–$1.4$, with good consistency for the two charges. Since

Table 4 Comparison between the ratios of figures of merit and the ratios of maximum heat loads for water, methanol and acetone

	N_w/N_m	N_w/N_a	N_m/N_a	N_w/N_m
	5.7	6.7	1.2	5.7
$v = 0.8$ ml	$q_{max,w}/N_m$	$q_{max,w}/q_{max,a}$	$q_{max,m}/q_{max,a}$	$q_{max,w}/N_m$
$v = 0.9$ ml	3.9	5.4	1.4	3.9

$\cos \theta$ should be smaller for water, the values of $q_{max,w}/q_{max,m}$ and $q_{max,w}/q_{max,a}$ are expected to be smaller than N_w/N_m and N_w/N_a, respectively. Therefore, the experimental data agree reasonably well with Eq. 3. It is noted that the effects of fluid charge and variation of the liquid layer thickness in the evaporator are not covered by the figure of merit.

As for $R_{e,min}$, Figs. 24 and 25 show different values for different fluids. It can be seen in Fig. 24 that water presents the lowest $R_{e,min}$s and acetone presents the highest. For better understanding, visualization of the liquid layer thickness in the evaporator is helpful. Figures 27, 28, and 29 show the visualization images along with their schematic side views for the three working fluids at the center of the heated zone. These images correspond to $R_{e,min}$ at $v = 0.8$ ml. In these conditions, the liquids at the front end (away from the condenser) are sustained in the lower layer. A slight increase of heat load would lead to local dryout at the front end. The total thickness of the sintered $100 + 200$ mesh wick is 0.26 mm. In Fig. 29, most of the upper 100 mesh screen, as well as the whole lower layer, is flooded in acetone. Although weak boiling occurred in this condition, the meniscus interlines appear unchanged by the slow up-and-down liquid motion. Once the interlines are identified in the top views shown in Figs. 27a, 28a and 29a, the vertical position of the interlines can be roughly located according to the side-view geometry of the mesh wick shown in Figs. 27b, 28b and 29b. According to the interline positions, $\delta_{w,eff}$ for acetone are roughly estimated as ~ 0.20–0.23 mm. In Fig. 28, the interline positions are slightly lower than those in Fig. 29, indicating a slightly thinner $\delta_{w,eff}$ of ~ 0.17–0.20 mm. In Fig. 27, most of the upper layer and the tops of the lower layer are exposed. Accordingly, $\delta_{w,eff}$ may lie between 0.08 and 0.15 mm. Inserting these $\delta_{w,eff}$ values into Eq. 4 with $k_{w,eff} = 13$ W/mK and $A_e = 1.21$ cm^2, we obtain $R_{e,min}$s of 0.13–0.14, 0.11–0.13 and 0.05–0.10 K/W for acetone, methanol and water, respectively. The experimental $R_{e,min}$s measured at the evaporator center are 0.11–0.12, 0.085–0.10 and 0.075–0.08 K/W for acetone, methanol and water, respectively (Figs. 24 and 26). Considering the uncertainties in R_e measurement and in estimating $\delta_{w,eff}$ from the images, the agreement between the approximate calculations by Eq. 4 and the experiments is remarkable. Therefore, the higher evaporator resistances for acetone can be mainly attributed to the larger average layer thicknesses in the evaporator. In fact, a significant amount of acetone is observed to accumulate at the condenser. This can be explained in that the low surface tension of acetone provides a low capillary pressure to draw the liquid from the condenser to the evaporator. In contrast, least accumulation is observed for water at the condenser with its high capillary pressures. Another

Fig. 27 **a** Top-view image
and **b** side-view schematic
plot of the evaporator center
for water at the minimum
evaporator resistance
with $v = 0.8$ ml and
$q = 70$ W/cm^2

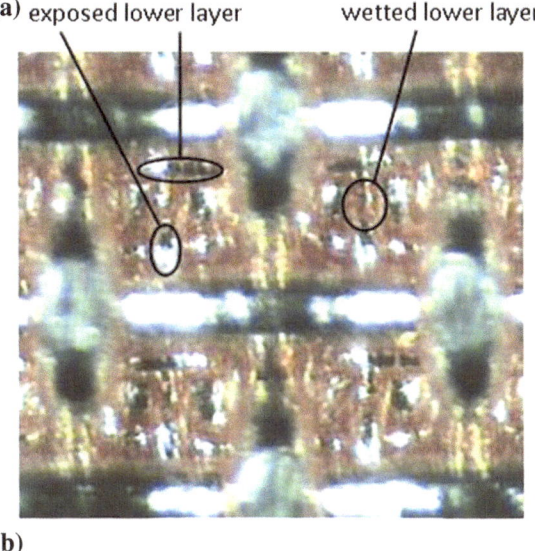

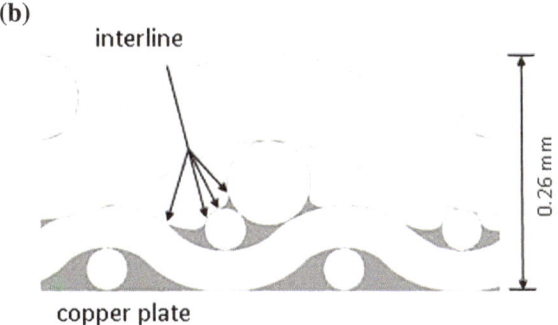

possible factor is the thermal conductivity of the liquids. From the analysis of
Ranjan et al. [10], the thin-film and non-thin-film resistances may occupy a portion
of the total evaporator resistance. Consequently, the fact that $k_{l,w} > k_{l,m} > k_{l,a}$
(Table 3) may be partly responsible for the different $R_{e,min}$s.

Figure 30 shows the superheats versus heat flux for different working fluids. In
spite of the large superheats for water, nucleation is absent. Even in a sintered
irregular-powder wick with abundant of nucleation sites, nucleate boiling is absent
for water up to a superheat of 13.7 K. For most of the cases, the superheat first
increases with the heat flux to a maximum value and then drops to a certain extent
until the onset of dryout. Since nucleate boiling is absent for water, the drop of
superheat is due to the thinning of the liquid layer. The very weak boiling of
methanol is not expected to affect the superheat values and the T1 readings only
show noise-level fluctuations. The drop of superheat near the q_{max} for methanol
should be mainly due to liquid layer thinning. For acetone, which exhibits slightly
stronger nucleation and weak fluctuations up to ±0.5 K in T1 near the q_{max}s

Fig. 28 a Top-view image
and **b** side-view schematic
plot of the evaporator center
for methanol at the minimum
evaporator resistance
with $v = 0.8$ ml and
$q = 22$ W/cm^2

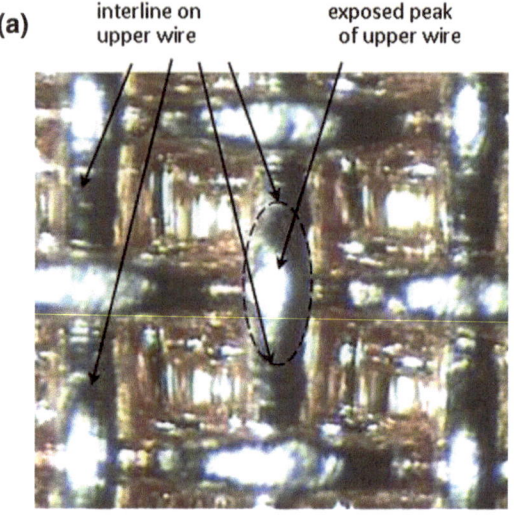

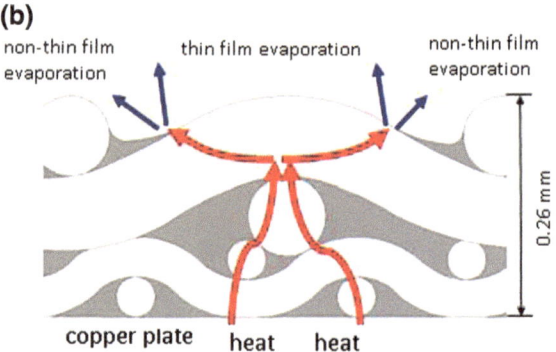

(cf. Fig. 24c), averaged superheat data are used in the plots. Obviously, the fluctuations result from the intermittent heat transfer enhancement associated with boiling. For acetone, the drastic drop of superheat near the maximum heat load may result from both liquid layer thinning and heat transfer enhancement by the nucleate boiling. The threshold superheats for nucleation of methanol and acetone are only about 1.3–4 K in the present mesh wick.

One thing to be addressed is the large differences of the threshold superheat between water and the other two fluids. For water, we have suggested that the absence of nucleation is because phase equilibrium could not prevail in the local evaporation region of an operating heat pipe, as the vapor above the evaporator rushes to the condenser. Thus, a higher superheat at the wick bottom is needed. For fluids of smaller surface tension, such as acetone and methanol, nucleate boiling tends to occur with less superheat. This is similar to the dependence on surface tension for heterogeneous pool boiling [19]. Nonetheless, their threshold

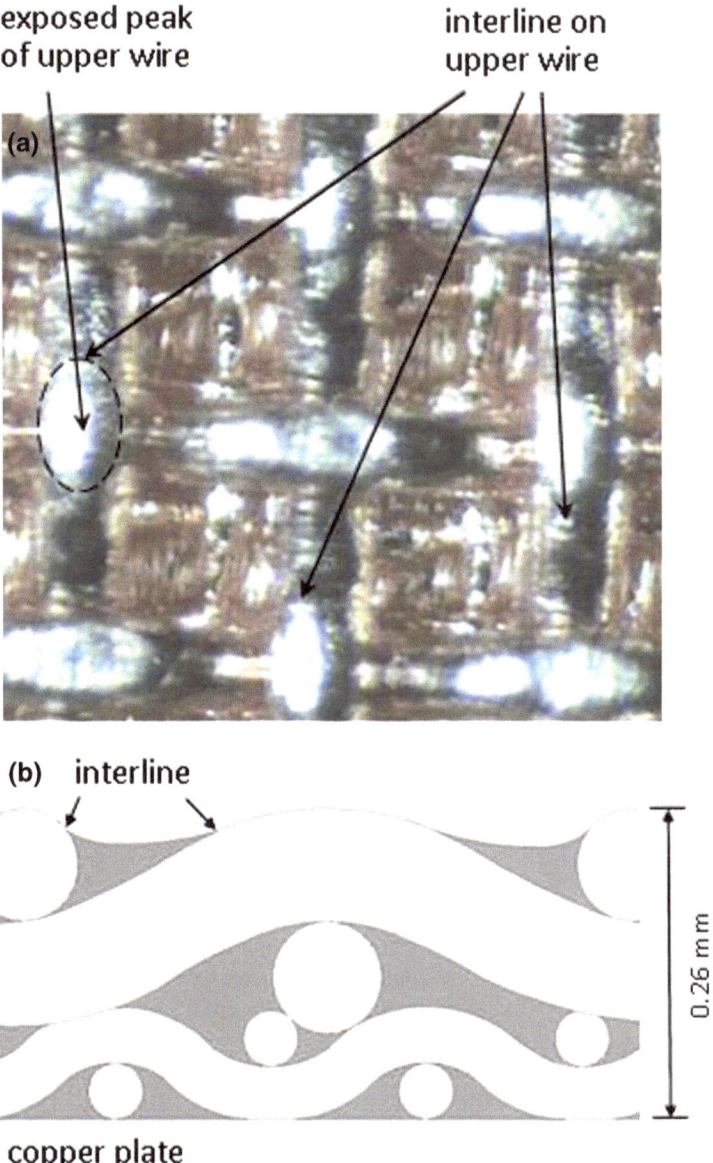

Fig. 29 a Top-view image and **b** side-view schematic plot of the evaporator center for acetone at the minimum evaporator resistance with $v = 0.8$ ml and $q = 16$ W/cm^2

superheats in the mesh evaporator of an operating heat pipe are still larger than the very small values observed for conditions with the liquid surface at phase equilibrium.

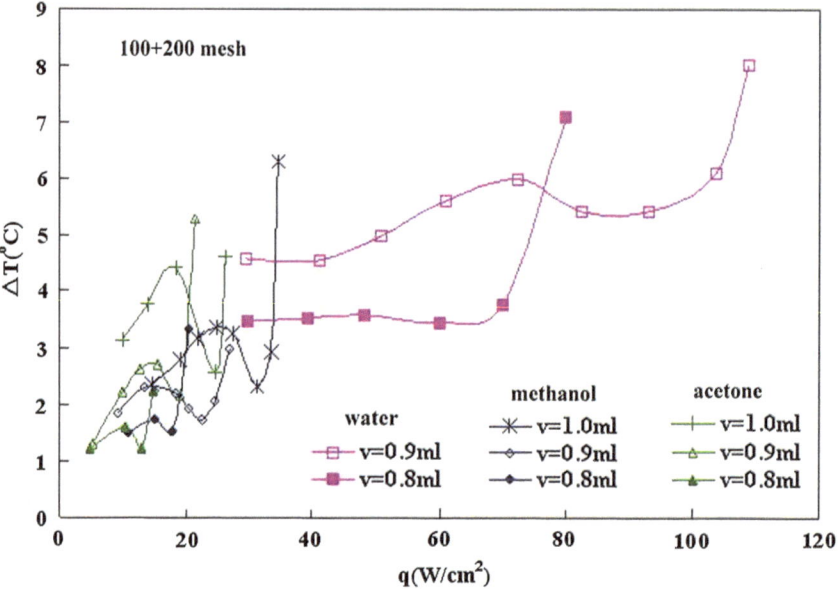

Fig. 30 Superheats at the evaporator versus heat flux for water, methanol, and acetone at the same charges, **a** $v = 0.9$ ml and **b** $v = 0.8$ ml

3.3 Effects of Surface Wettability

Figure 31 presents the static contact angles of 5 µl water drops measured at different elapsed times in air at 25–30 °C. These data were collected over an experiment period of about three months with wide variation of humidity. The somewhat scattered data points may result from the fact that the contact angle growth rate is influenced by both environmental temperature and humidity. Nonetheless, it is shown that the contact angle increases roughly linearly with the elapsed time, from 9° to about 40° during 3 h. Figure 32 illustrates some typical sessile water drops at different elapsed times.

Figure 33 compares the evaporative resistances (R_e) versus heat flux for water at a charge of $v = 0.9$ ml with respect to the static contact angle, θ_0. The values of R_e are obviously smaller for a larger θ_0 before the maximum heat loads. Visualization reveals earlier water film recession in the evaporator for a larger θ_0, apparently resulting from the lower wettability. However, the minimum values of R_e are about the same for various θ_0. This point will be discussed later. In addition, the maximum heat load, beyond which local dryout appears, decreases with increasing θ_0, or decreasing surface wettability. According to Eq. 1, the capillary force provided by the evaporating wick is proportional to $\cos\theta$, where θ is the apparent contact angle at the meniscus contact lines. Since the wick is identical, the maximum heat loads for different tests should be proportional to $\cos\theta$. It is

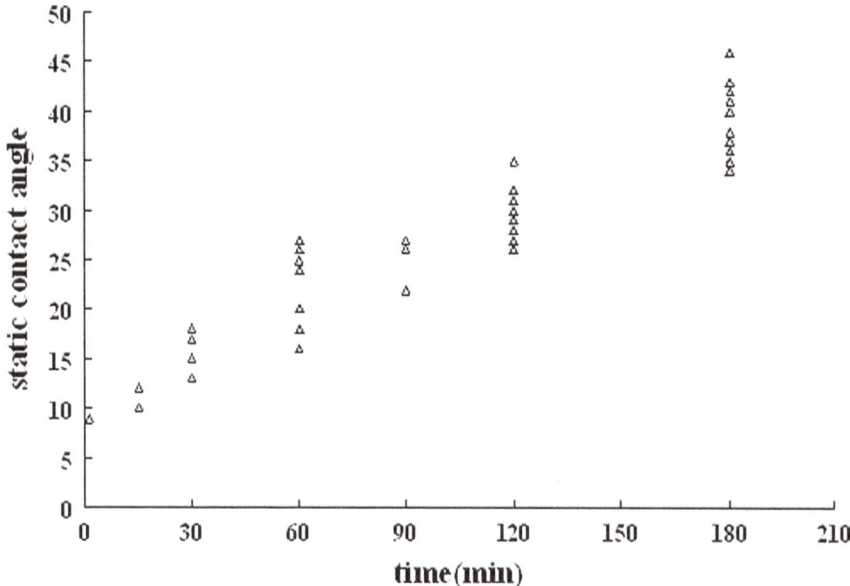

Fig. 31 Contact angles of 5 μl sessile water drops on a smooth copper surface at different elapsed times

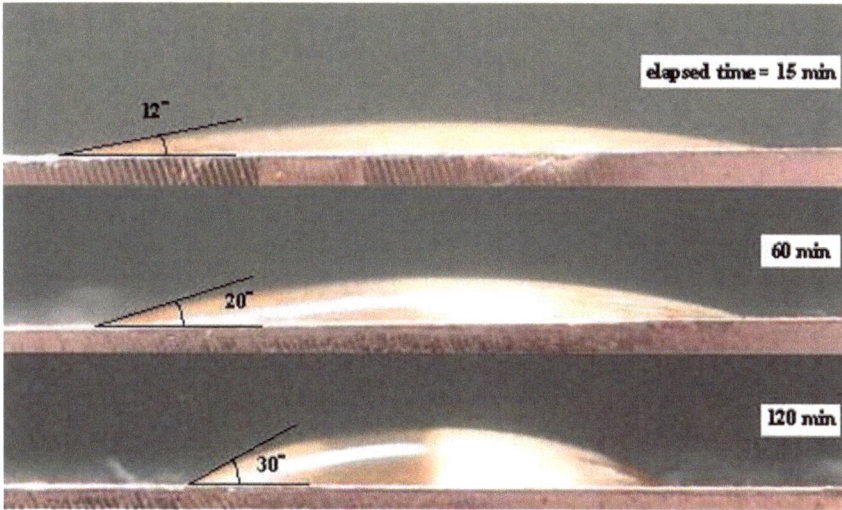

Fig. 32 Typical sessile water drops on a smooth copper surface at different elapsed times

known that the apparent contact angles θ under evaporation are larger than the static contact angles θ_0, with larger deviation for stronger evaporation [18]. To exhibit the relation between maximum heat load and contact angle, we plot $q_{max,\theta}$/

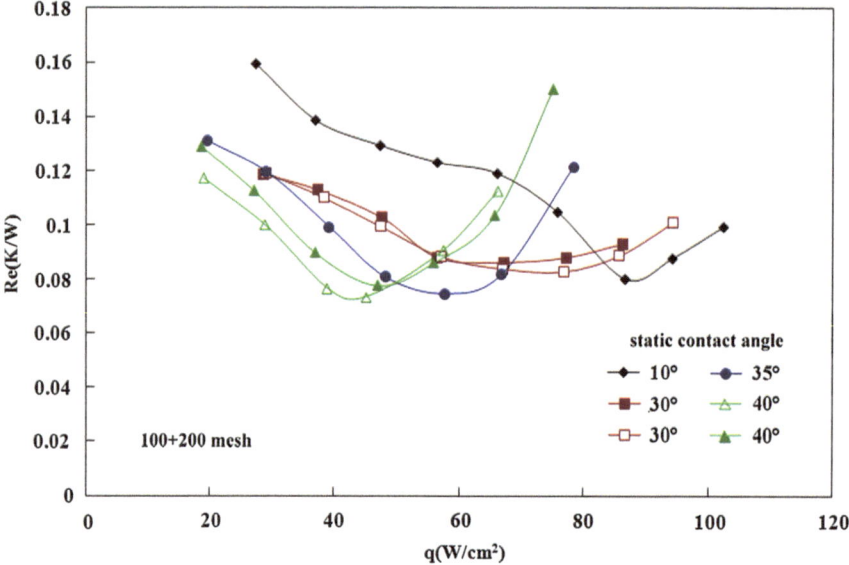

Fig. 33 Evaporative resistances versus heat flux for water with copper surfaces presenting different static contact angles, $v = 0.9$ ml

$q_{max, \, 10°}$ versus θ_0 in Fig. 34, along with the curve of $\cos\theta_0$ for reference. For $\theta_0 > 30°$, $q_{max,\theta}$ drops more than the reduction associated with $\cos\theta_0$, implying that the larger θ_0 is not the only reason for q_{max} reduction. Additional reason may be attributed to the amplification in the apparent contact angle θ. From the microscopic view for thin film evaporation, the attraction force for liquid flow comes from both the disjoining pressure and the capillary pressure, while viscous friction is the main resistance force. According to the theoretical comparison between water and other non-polar fluids [20] and between polar and artificial non-polar water [21], a high polarity induces a much thicker evaporating thin film as well as a stronger disjoining pressure. The much thicker film would result in less friction, although water has a higher viscosity than other fluids. In the present cases with a lowered copper surface energy but unaltered polarity, we could expect a weaker disjoining pressure and a thinner film, similar to the effect of reducing polarity. The thinner film would render in higher friction, while the weakened disjoining pressure would draw a smaller amount of water into the evaporating thin film region. Both are unfavorable to the maximum heat load. Macroscopically, enhanced amplification in the apparent contact angle θ would appear for larger θ_0 as a result of evaporation, weakened disjoining pressure, and increased friction.

As far as the minimum R_e is concerned, it is readily known that R_e is dominated by $\delta_{w,eff}$ (Eq. 4), as the surface evaporation on the menisci presents a negligible thermal resistance. With an identical wick, the minimum R_e appears independent of surface wettability.

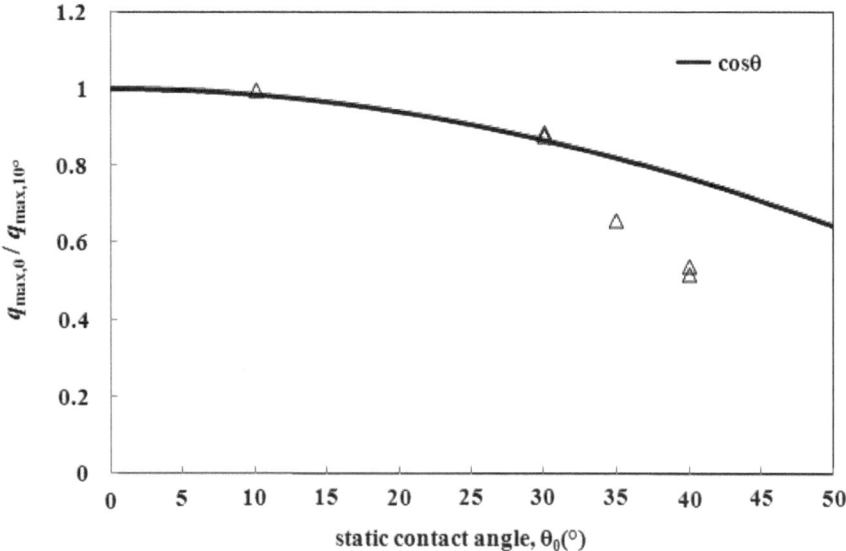

Fig. 34 $q_{max,\theta}/q_{max,10°}$ for water with copper surfaces presenting different static contact angles, $v = 0.9$ ml

It can be stated for copper/water heat pipes or vapor chambers that their maximum heat loads would degrade with the exposure time of the sintered wick in air during the manufacturing process. The maximum heat loads may be halved with an exposure time of 3 h. However, the minimum R_e would hardly be affected.

In Figs. 35 and 36, the evaporation performances with the copper wick exposed in air for different elapsed times are shown for methanol and acetone, respectively. As has been noted, even though the copper surface energy decreases with the elapsed time, reflected by the increasing θ_0 for water, the surfaces are wetted by methanol and acetone due to their low surface tensions. Interestingly, here we find both the maximum heat loads and the evaporative resistances nearly unaffected by the variation of copper surface energy. Only slightly lower evaporative resistances appear prior to the maximum heat loads. Also, the maximum heat loads for increasing θ_0 are nearly unaffected. Since both methanol and acetone have a polarity index a half of that of water, the film would be thinner and the disjoining pressure would be weaker [20, 21]. For these fluids, the relative role of the friction force grows with reduced film thickness. Kim and Wayner [18] indicated that, for completely wetting liquids, friction force is the controlling factor in the thin evaporating film region where the apparent contact angle θ is determined. In other words, the thin film evaporation process associated with a less-polar fluid should be less sensitive to disjoining pressure variation than for a highly polar fluid. Shortly speaking, the evaporation-enlarged θ from $\theta_0 = 0°$ was insensitive to the surface energy variation that $\cos \theta$ and Q_{max} (Eq. 3) are similar.

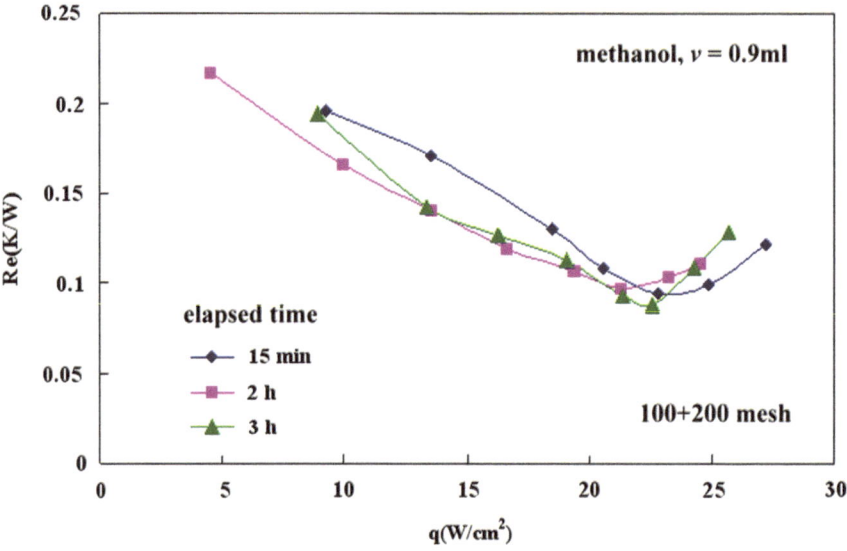

Fig. 35 Evaporative resistances versus heat flux for methanol with copper surfaces exposed in air for different elapsed times, $v = 0.9$ ml

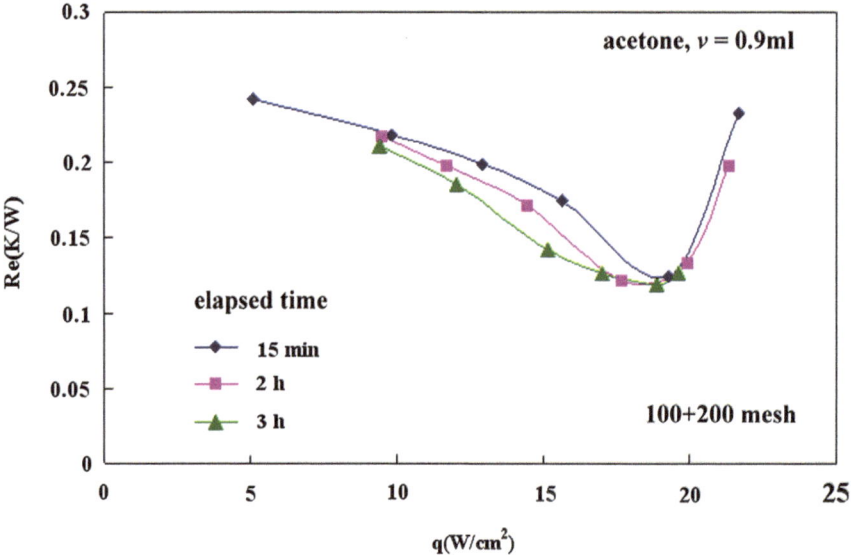

Fig. 36 Evaporative resistances versus heat flux for acetone with copper surfaces exposed in air for different elapsed times, $v = 0.9$ ml

4 Conclusions

Combined evaporator resistance measurement and visualization have been conducted for operating flat-plate heat pipes. First examined are different wick types, including sintered multi-layer copper-mesh wicks and sintered copper-powder wicks, using water as the working fluid. Then, different working fluids, including water, methanol and acetone, are investigated using an identical two-layer mesh wick under various copper surface wettabilities. The following conclusions are obtained:

1. Quiescent surface evaporation prevails for sintered-mesh or sintered-powder wicks working with water. For either irregular powders or spherical powders, no nucleate boiling is observed up to heat flux larger than 100 W/cm^2 in spite of the abundant nucleation sites. The maximum superheat of 13.7 °C obtained in this work is insufficient to invoke nucleate boiling.
2. With increasing heat load, the water layer recedes to form corrugated menisci within the wick. The evaporator resistance decreases until local dryout occurs. Local dryout begins from the front end and expands across the evaporator, leading to gradual rise of the evaporator resistance.
3. According to the visualization for two-layer mesh wicks, the minimum evaporator resistance exists when a thin water layer is sustained in the bottom mesh layer.
4. Fine pores at the wick bottom help sustain a thin water layer under large heat loads. Smaller evaporator resistances as well as large maximum heat loads could be reached.
5. Although the strong capillarity of irregular-powder wick favors in evaporation performance, its low permeability retards the back flow of the condensed water when the heat pipe is homogeneously laid with irregular powders. However, this can be compensated by adopting a larger charge and/or a thicker wick.
6. The maximum heat loads for water are far greater than those of methanol and acetone. These values for different working fluids correlate well with their figures of merit subject to modification for different apparent contact angles.
7. The minimum evaporator resistances for the three working fluids differ slightly, with water the lowest and acetone the highest. This mainly results from the visualized fact that the evaporating water layers are thinnest and the acetone layers are thickest. However, it may also be related with the fact that water has the largest thermal conductivity and acetone has the smallest.
8. While quiet surface evaporation prevails for water in the present test conditions, weak and slow nucleate boiling occurs for acetone and methanol near the maximum heat loads but is again suppressed beyond them. Due to the weakness of the nucleation, the maximum heat loads should be capillary-limited.
9. A superheat of about 1.3–4 K at the wick bottom is sufficient to activate nucleate boiling of acetone and methanol due to their low surface tensions.

10. For copper/water heat pipes or vapor chambers, their critical heat loads would degrade with the exposure time of the sintered wick in air during the manufacturing process. For an exposure time of 3 h, the critical heat loads could be halved. However, the minimum evaporative resistances would hardly be affected.

11. For methanol or acetone, a smooth copper surface would still be wetted even after an exposure time of 3 h. Both the critical heat loads and the minimum evaporative resistances are nearly unaffected.

References

A. Faghri, *Heat Pipe Science and Technology* (Taylor and Francis, London, 1995)

C. Li, G.P. Peterson, Y. Wang, Evaporation/boiling in thin capillary wicks (I)—wick thickness effects. ASME J. Heat Transf. **128**, 1312–1319 (2006)

C. Li, G.P. Peterson, Evaporation/boiling in thin capillary wicks (II)—effects of volumetric porosity and mesh size. ASME J. Heat Transf. **128**, 1320–1328 (2006)

A. Brautsch, P.A. Kew, Examination and visualization of heat transfer processes during evaporation in capillary porous structures. Appl. Therm. Eng. **22**, 815–824 (2002)

S.-C. Wong, Y.-H. Kao, Visualization and performance measurement of operating meshed-wick heat pipes. Int. J. Heat Mass Transf. **51**, 4249–4259 (2008)

S.W. Chi, *Heat Pipe Theory and Practice* (McGraw-Hill, New York, 1976)

R. Kempers, A.J. Robinson, D. Ewing, C.Y. Ching, Characterization of evaporator and condenser thermal resistances of a screen mesh wicked heat pipe. Int. J. Heat Mass Transf. **51**, 6039–6046 (2008)

J.-Y. Chang, R.S. Prasher, S. Prstic, P. Cheng, H.B. Ma, Evaporative thermal performance of vapor chambers under nonuniform heating conditions. ASME J. Heat Transf. **130**, 121501 (2008)

G.S. Hwang, Y. Nam, E. Fleming, P. Dussinger, Y.S. Ju, M. Kaviany, Multi-artery heat pipe spreader: experiment. Int. J. Heat Mass Transf. **53**, 2662–2669 (2010)

R. Ranjan, J.Y. Murthy, S.V. Garimella, Analysis of the wicking and thin-film evaporation characteristics of microstructures. ASME J. Heat Transf. **131**, 101001 (2009)

K.K. Bodla, J.Y. Murthy, S.V. Garimella, Evaporation analysis in sintered wick microstructures. Int. J. Heat Mass Transf. **61**, 729–741 (2013)

J.-H. Liou, C.-W. Chang, C. Chao, S.-C. Wong, Visualization and thermal resistance measurement for the sintered mesh-wick evaporator in operating flat-plate heat pipes. Int. J. Heat Mass Transf. **53**, 1498–1506 (2010)

S.-C. Wong, J.-H. Liou, C.-W. Chang, Evaporation resistance measurement and visualization for sintered copper-powder evaporator in operating flat-plate heat pipes. Int. J. Heat Mass Transf. **53**, 3792–3798 (2010)

S.-C. Wong, Y.-C. Lin, J.-H. Liou, Visualization and evaporator resistance measurement in heat pipes charged with water, methanol or acetone. Int. J. Therm. Sci. **52**, 154–160 (2012)

S.-C. Wong, Y.-C. Lin, Effect of copper surface wettability on the evaporation performance of heat pipes. Int. J. Heat Mass Transf. **54**, 3921–3926 (2011)

S.-C. Wong, K.-C. Hsieh, J.-D. Wu, W.-L. Han, A novel vapor chamber and its performance. Int. J. Heat Mass Transf. **53**, 2377–2384 (2010)

T. Semenic, I. Catton, Experimental study of biporous wicks for high heat flux applications. Int. J. Heat Mass Transf. **52**, 5113–5121 (2009)

I.Y. Kim, P.C. Wayner, Shape of an evaporating completely wetting extended meniscus. J. Thermophys. Heat Transf. **10**, 320–325 (1996)

V.P. Carey, *Liquid-Vapor Phase-Change Phenomena*, 2nd edn. (Taylor and Francis, London, 2007)

W. Qu, T. Ma, Effects of the polarity of working fluids on vapor-liquid flow and heat transfer characteristics in a capillary. Microscale Thermophys. Eng. **6**, 175–190 (2002)

S.-K. Wee, K.D. Kihm, K.P. Hallinan, Effects of the liquid polarity and the wall slip on the heat and mass transport characteristics of the micro-scale evaporating transition film. Int. J. Heat Mass Transf. **48**, 265–278 (2005)